JESSICA HAGY'S

AETUI

(A to E)

PENTAGRAM POEMS

FROM

INSIDE THE CASTLE

A small press specializing in works of literature from the "expanded field"
Lawrence, Kansas

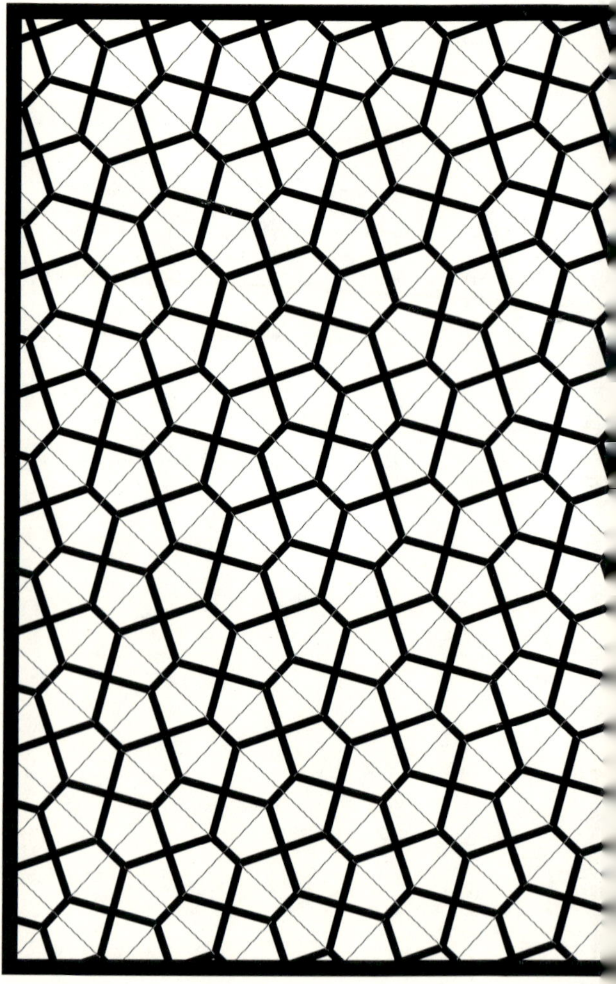

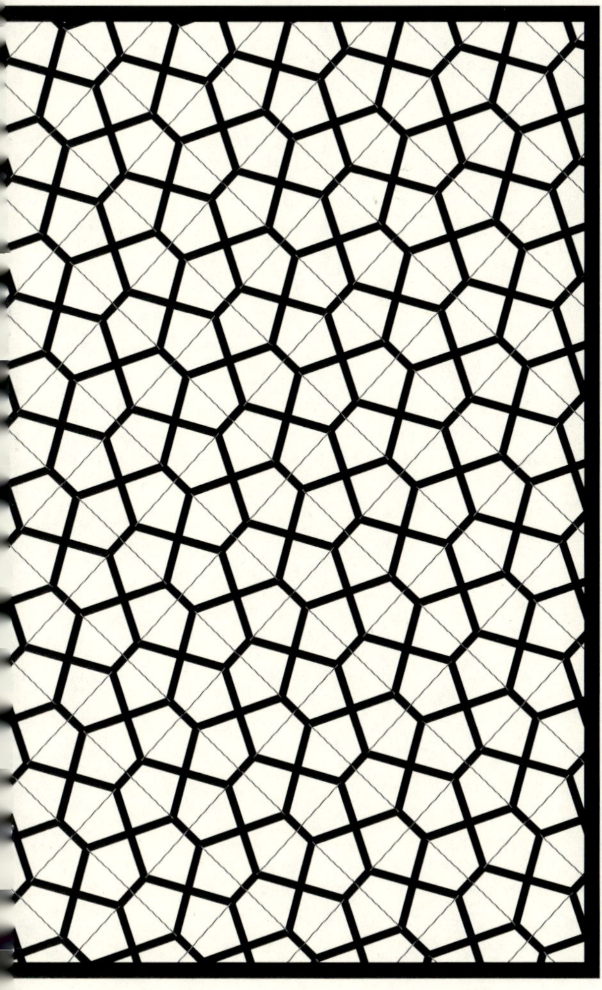

JESSICA HAGY'S

AETUI

***Figure 1*: Super-human powers**

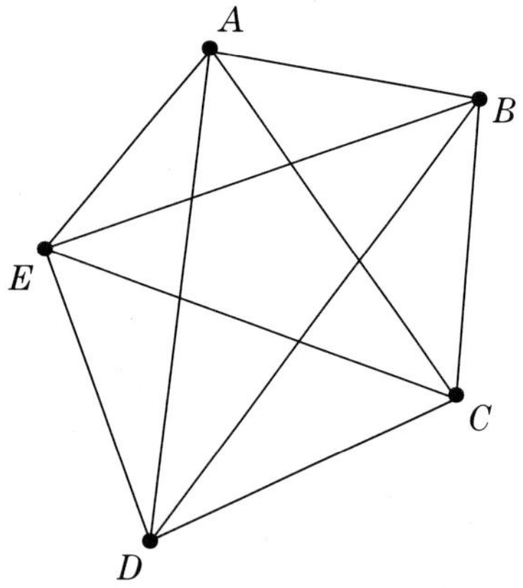

A = Flying
B = Invisibility
C = Time travel
D = Immortality
E = Empathy

$\overline{AB}$ = Engage cloaking device

$\overline{AC}$ = Crossing the IDL

$\overline{AD}$ = Vampire bats

$\overline{AE}$ = Ceding the armrest

$\overline{BC}$ = Ghosts

$\overline{BD}$ = The concept of a soul

$\overline{BE}$ = Anonymous gifts

$\overline{CD}$ = Wormholes

$\overline{CE}$ = Killing Hitler

$\overline{DE}$ = Mentoring someone

***Figure 2*: Any reason to celebrate**

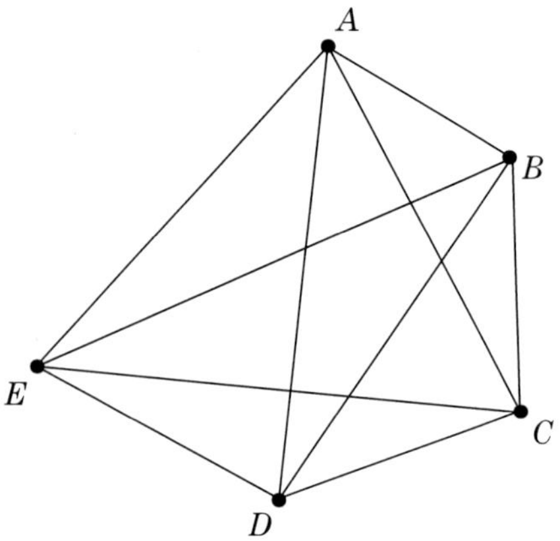

A = Christmas

B = Easter

C = Halloween

D = Thanksgiving

E = Valentine's Day

$\overline{AB}$ = Chocolate

$\overline{AC}$ = Scrooge's ghosts

$\overline{AD}$ = Thank-you notes

$\overline{AE}$ = Mommy kissing Santa

$\overline{BC}$ = Zombies

$\overline{BD}$ = No good presents

$\overline{BE}$ = So many bunnies

$\overline{CD}$ = Yay, harvests

$\overline{CE}$ = Stalkers

$\overline{DE}$ = Not being alone

***Figure 3*: Witch skills**

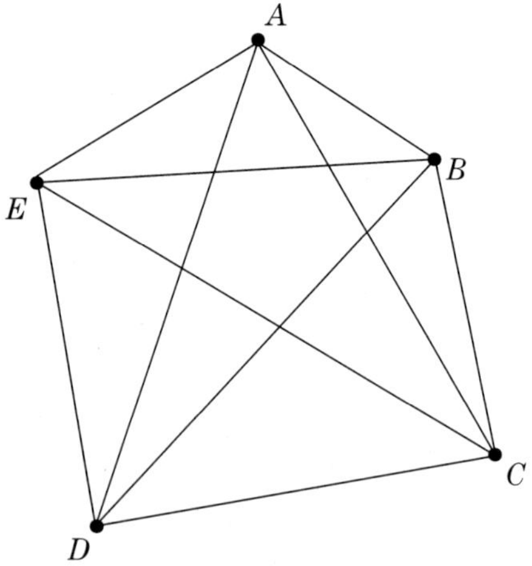

A = Cursing

B = Flying

C = Transfiguration

D = Conjuring

E = Divining

$\overline{AB}$ = Aisle seats

$\overline{AC}$ = Ill-fitting clothes

$\overline{AD}$ = Amateur magicians

$\overline{AE}$ = Leaky plastic 8-ball

$\overline{BC}$ = Brooms

$\overline{BD}$ = Hatching eggs

$\overline{BE}$ = Aerodynamics

$\overline{CD}$ = Extreme make-overs

$\overline{CE}$ = Watching your parents age

$\overline{DE}$ = Tomorrow's to-do list

Figure 4: The five horsemen of the Apocalypse

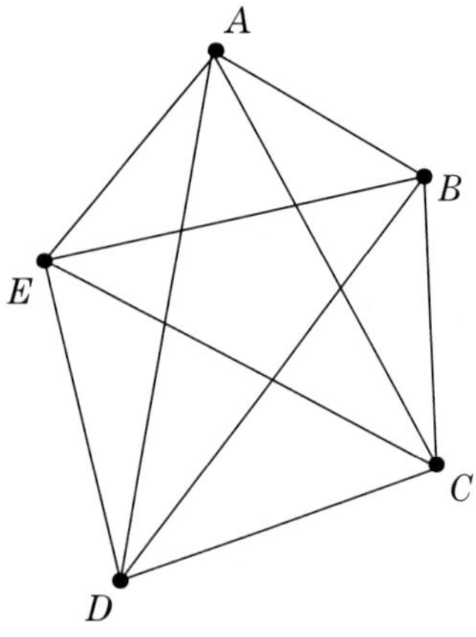

A = Conquest

B = War

C = Famine

D = Death

E = Ratings

$\overline{AB}$ = Imperialism

$\overline{AC}$ = Refugees

$\overline{AD}$ = Genocide

$\overline{AE}$ = Politics

$\overline{BC}$ = Scorched earth

$\overline{BD}$ = Always

$\overline{BE}$ = Shock & Awe

$\overline{CD}$ = Starvation

$\overline{CE}$ = Star-studded telethons

$\overline{DE}$ = Live executions

***Figure 5*: Strength through conditioning**

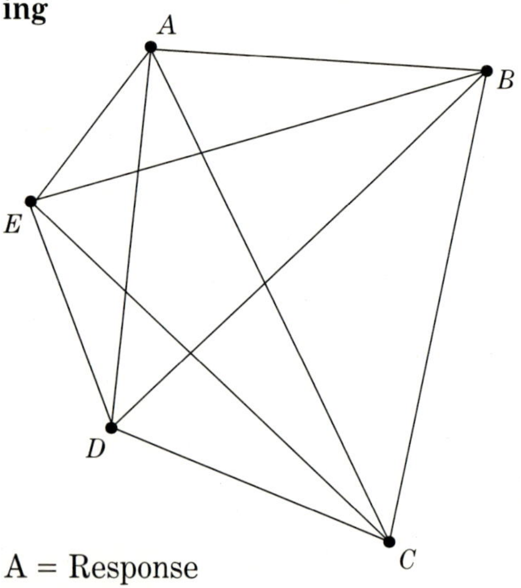

A = Response

B = Hair

C = Air

D = Behavioral

E = Athletic

$\overline{AB}$ = Don't ask to touch

$\overline{AC}$ = Adjust the thermostat

$\overline{AD}$ = Please & thank you

$\overline{AE}$ = Run when the gun goes off

$\overline{BC}$ = Blow dryer

$\overline{BD}$ = Trichotillomania

$\overline{BE}$ = Ponytail

$\overline{CD}$ = Heavy sigh

$\overline{CE}$ = Nikes

$\overline{DE}$ = Never skipping practice

***Figure 6*: Fragments of thought**

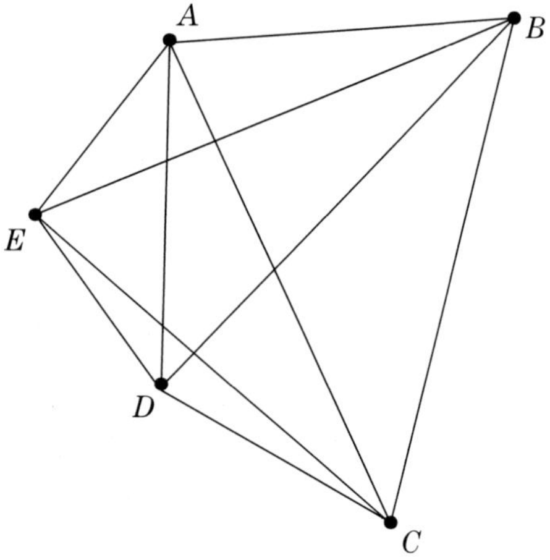

A = Noun
B = Verb
C = Modifier
D = Interjection
E = Conjunction

$\overline{AB}$ = You're

$\overline{AC}$ = Beautiful you

$\overline{AD}$ = My god

$\overline{AE}$ = And you

$\overline{BC}$ = Are so

$\overline{BD}$ = Are, oh my god

$\overline{BE}$ = Are and

$\overline{CD}$ = Beautiful, oh my god

$\overline{CE}$ = Beautiful and

$\overline{DE}$ = Oh my god, and

***Figure 7*: Evangelical topics**

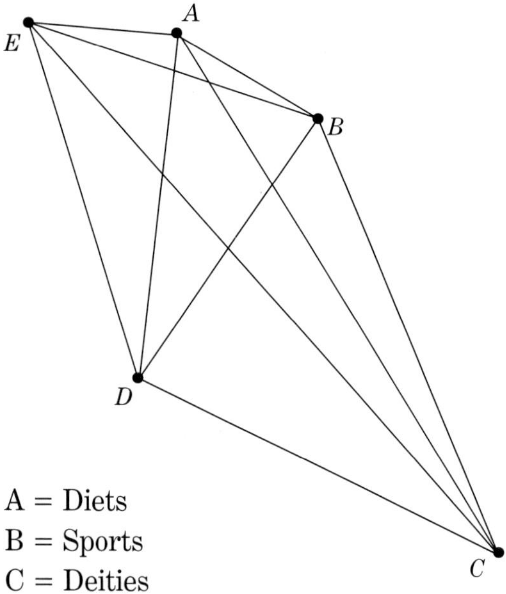

A = Diets
B = Sports
C = Deities
D = Pyramid schemes
E = Media

$\overline{AB}$ = Getting in shape

$\overline{AC}$ = Fasting

$\overline{AD}$ = Juice cleanses

$\overline{AE}$ = Unreal expectations

$\overline{BC}$ = Gold medalists

$\overline{BD}$ = Childhood athletics

$\overline{BE}$ = Corporate sponsorships

$\overline{CD}$ = Door-to-door believers

$\overline{CE}$ = The Jesus Channel

$\overline{DE}$ = Broke moms on Facebook

Figure 8: On the striking of gold

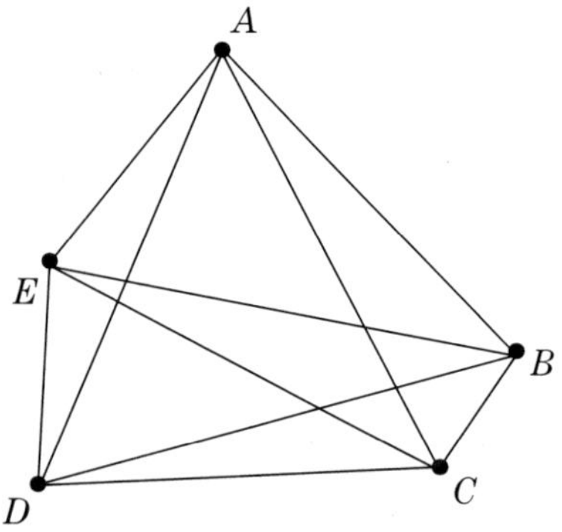

A = Alchemy
B = Business
C = Lottery
D = Marriage
E = Mining

$\overline{AB}$ = Branding

$\overline{AC}$ = Beating the odds

$\overline{AD}$ = Chemistry

$\overline{AE}$ = Klondike (1886-1899)

$\overline{BC}$ = Cold-calling

$\overline{BD}$ = The politics of royalty

$\overline{BE}$ = Data analysis

$\overline{CD}$ = Awkward arrangements

$\overline{CE}$ = Veruca's golden ticket

$\overline{DE}$ = Mail-order brides

Figure 9: Can't look away

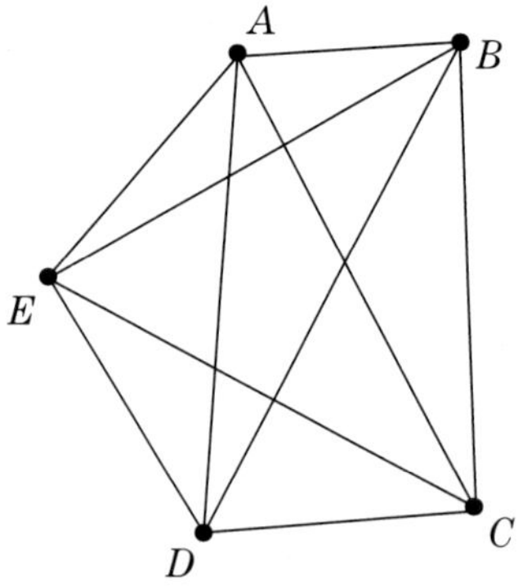

A = Novelty
B = Atrocity
C = Beauty
D = Complexity
E = Screens

$\overline{AB}$ = Train wrecks

$\overline{AC}$ = Models

$\overline{AD}$ = Puzzles

$\overline{AE}$ = News

$\overline{BC}$ = Rising from ashes

$\overline{BD}$ = Nazi bureaucracy

$\overline{BE}$ = Terrorist chat rooms

$\overline{CD}$ = Nature

$\overline{CE}$ = Photoshop

$\overline{DE}$ = Dynamic networks

Figure 10: What delightful is

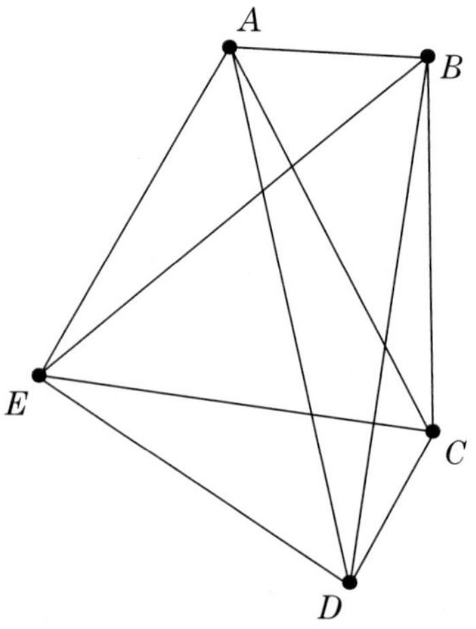

A = Gentle
B = Thoughtful
C = Surprising
D = Nourishing
E = Unserious

$\overline{AB}$ = Small kindnesses

$\overline{AC}$ = Hopeful words

$\overline{AD}$ = Good soups

$\overline{AE}$ = Friendly laughter

$\overline{BC}$ = Perfect gifts

$\overline{BD}$ = Needed advice

$\overline{BE}$ = Phone calls from old friends

$\overline{CD}$ = Spur-of-the-moment trips

$\overline{CE}$ = Compliments from strangers

$\overline{DE}$ = Paperback novels

Figure 11: Getting the clues together

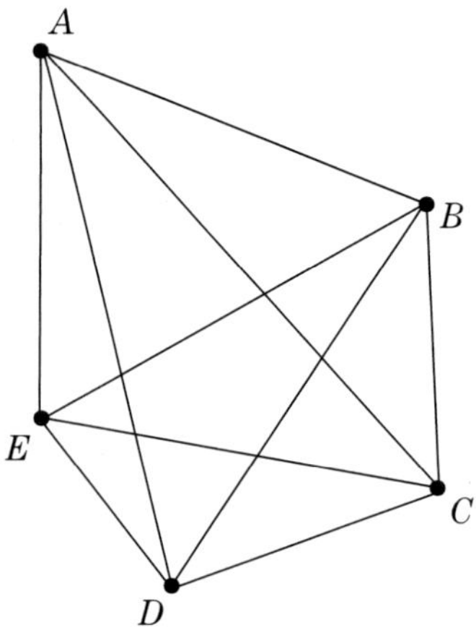

A = Who
B = What
C = Where
D = When
E = Why

$\overline{AB}$ = Col. Mustard with the knife

$\overline{AC}$ = Col. Mustard in the closet

$\overline{AD}$ = Col. Mustard last night

$\overline{AE}$ = Col. Mustard's secret lover

$\overline{BC}$ = Dagger to the heart

$\overline{BD}$ = Silenced before he could tell

$\overline{BE}$ = Death before outing

$\overline{CD}$ = Everywhere in the past

$\overline{CE}$ = In the closet for survival

$\overline{DE}$ = Last night fear defeated love

Figure 12: Talking around a problem

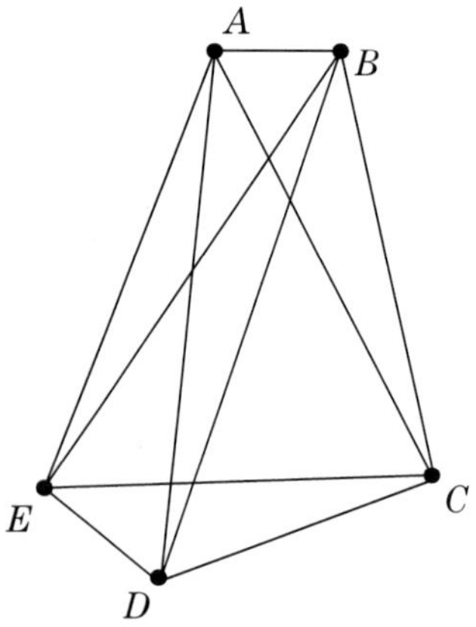

A = Before
B = During
C = Later
D = Instead
E = Perhaps

$\overline{AB}$ = Lukewarm hope

$\overline{AC}$ = Silent regret

$\overline{AD}$ = Daring to trust myself

$\overline{AE}$ = I knew

$\overline{BC}$ = Ironic loneliness

$\overline{BD}$ = Back out to safety

$\overline{BE}$ = Someone would object

$\overline{CD}$ = Live alone and fearlessly

$\overline{CE}$ = It will get better

$\overline{DE}$ = Someone will love me

***Figure 13*: Crossing genres**

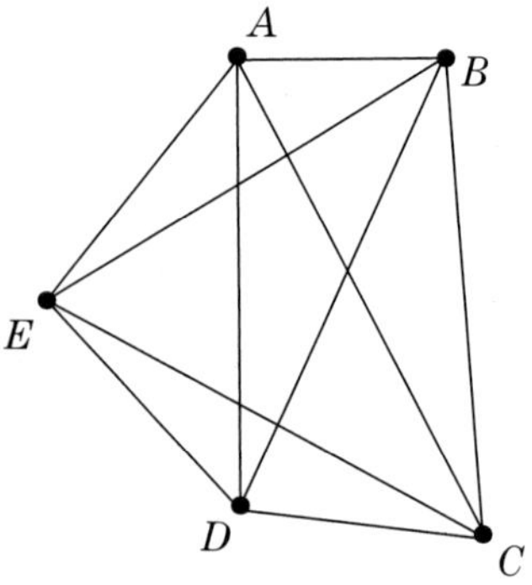

A = Drama
B = Mystery
C = Comedy
D = Romance
E = Science Fiction

$\overline{AB}$ = Who started it?

$\overline{AC}$ = Revealing Freudian slips

$\overline{AD}$ = On again, off again

$\overline{AE}$ = DNA tests

$\overline{BC}$ = Who farted?

$\overline{BD}$ = Blind dates

$\overline{BE}$ = Could that really happen?

$\overline{CD}$ = Friendship

$\overline{CE}$ = Sarcastic robots

$\overline{DE}$ = Sexy robots

Figure 14: Whatever floats your boat

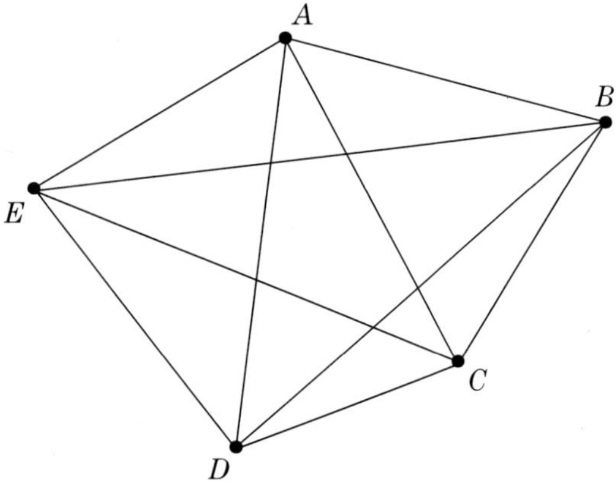

A = Voyeurism
B = Exhibitionism
C = Masochism
D = Sadism
E = Capitalism

$\overline{AB}$ = Open drapes

$\overline{AC}$ = Internet stalking your exes

$\overline{AD}$ = Rubbernecking

$\overline{AE}$ = Reality television

$\overline{BC}$ = Social media

$\overline{BD}$ = Revenge porn

$\overline{BE}$ = Stripping

$\overline{CD}$ = Suicide bombers

$\overline{CE}$ = Employees

$\overline{DE}$ = Employers

Figure 15: And you are?

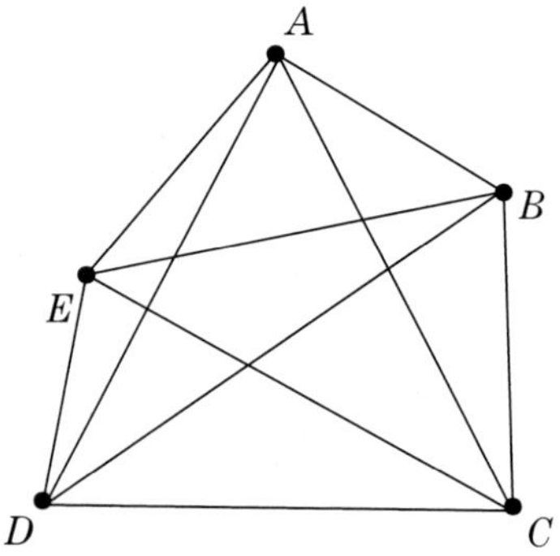

A = Animal
B = Vegetable
C = Mineral
D = Spiritual
E = Fictional

$\overline{AB}$ = Stew

$\overline{AC}$ = Bones

$\overline{AD}$ = Patronus

$\overline{AE}$ = Unicorn

$\overline{BC}$ = Petrified forests

$\overline{BD}$ = Ayahuasca

$\overline{BE}$ = Killer tomatoes

$\overline{CD}$ = Crystals

$\overline{CE}$ = Kryptonite

$\overline{DE}$ = Zeus & friends

***Figure 16*: Why'd you do that?**

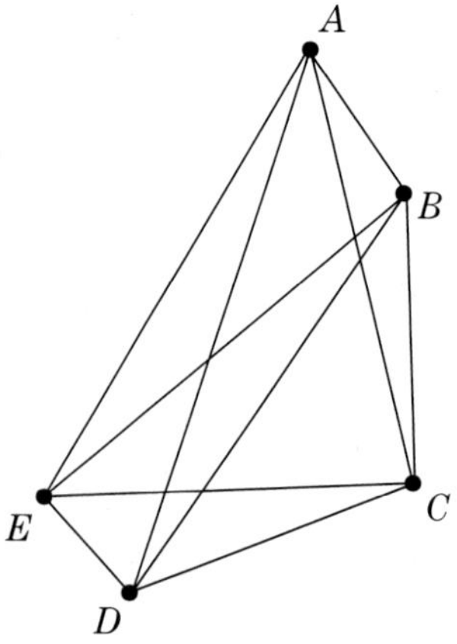

A = Habit
B = Ritual
C = Routine
D = Duty
E = Compulsion

$\overline{AB}$ = Gesundheit

$\overline{AC}$ = Double-check the locks

$\overline{AD}$ = Nun's hats

$\overline{AE}$ = Nail biting

$\overline{BC}$ = Bedtime prayers

$\overline{BD}$ = Sunday mass

$\overline{BE}$ = Triple-check the locks

$\overline{CD}$ = Commuting

$\overline{CE}$ = Snacking

$\overline{DE}$ = Voting

Figure 17: Basic crayons

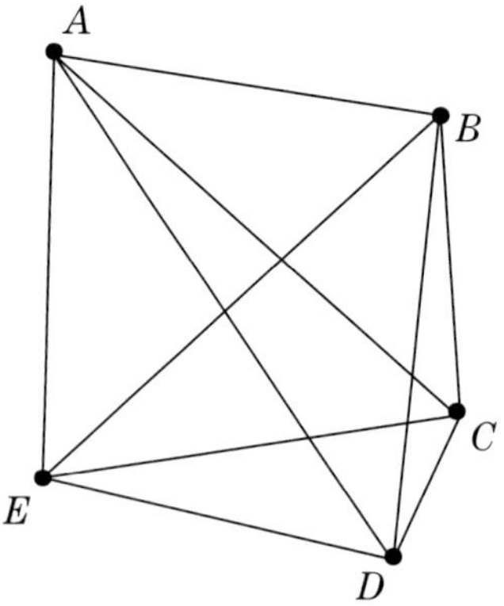

A = Red
B = Orange
C = Yellow
D = Green
E = Blue

$\overline{AB}$ = Campfire

$\overline{AC}$ = Fast food

$\overline{AD}$ = Christmas

$\overline{AE}$ = Teary eyes

$\overline{BC}$ = Smoggy morning

$\overline{BD}$ = Flag of Ireland

$\overline{BE}$ = Penicillin

$\overline{CD}$ = Chartreuse

$\overline{CE}$ = Regretfully timid

$\overline{DE}$ = Environmental researchers

Figure 18: Composing motivation

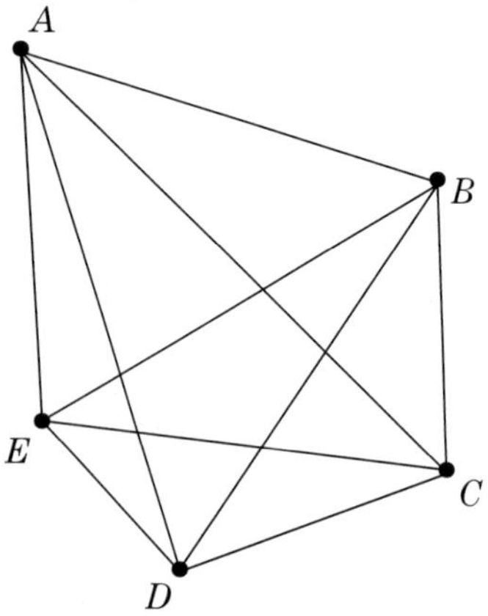

A = Could
B = Should
C = Might
D = Will
E = Can

$\overline{AB}$ = Do better more often

$\overline{AC}$ = Do better more often

$\overline{AD}$ = Do better more often

$\overline{AE}$ = Do better more often

$\overline{BC}$ = Do better more often

$\overline{BD}$ = Do better more often

$\overline{BE}$ = Do better more often

$\overline{CD}$ = Do better more often

$\overline{CE}$ = Do better more often

$\overline{DE}$ = Do better more often

Figure 19: Plot points in history

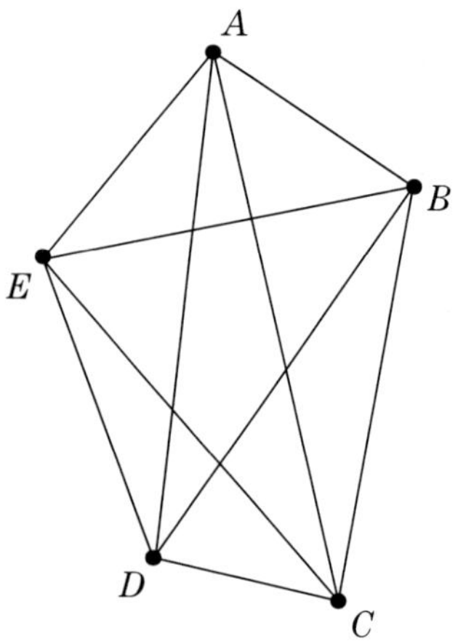

A = Germs

B = Guns

C = Steel

D = God

E = Babies

$\overline{AB}$ = Military quarantine

$\overline{AC}$ = Bloody swords

$\overline{AD}$ = Last rites

$\overline{AE}$ = Daycare

$\overline{BC}$ = Tanks

$\overline{BD}$ = Modern crusades

$\overline{BE}$ = Tragic headlines

$\overline{CD}$ = Excalibur

$\overline{CE}$ = Monuments

$\overline{DE}$ = Christmas

Figure 20: A chorus of voices

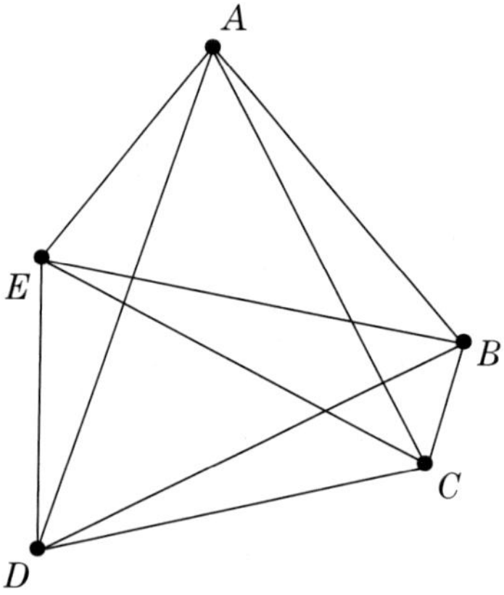

A = Imperative
B = Declarative
C = Interrogative
D = Exclamatory
E = Obfuscating

$\overline{AB}$ = I think you should eat shit.

$\overline{AC}$ = Eat that shit I told you to?

$\overline{AD}$ = Eat shit!

$\overline{AE}$ = Gormandize man-guano.

$\overline{BC}$ = You ate that shit?

$\overline{BD}$ = You ate that shit!

$\overline{BE}$ = Your ordure is served.

$\overline{CD}$ = How full of shit are you?

$\overline{CE}$ = Is the feculence saporous?

$\overline{DE}$ = Ingest defecation!

***Figure 21*: Narrating a hook-up**

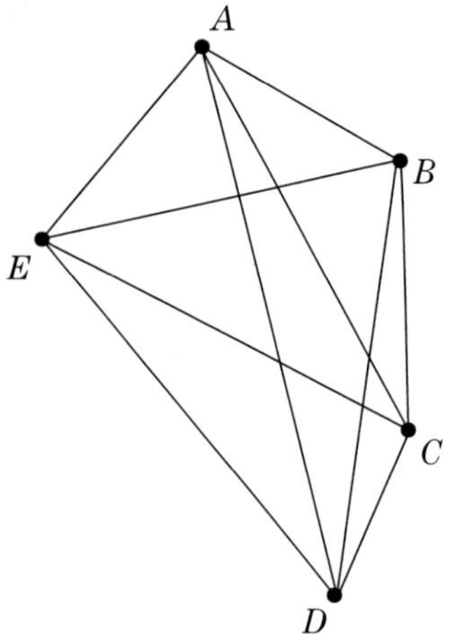

A = Exposition
B = Rising action
C = Climax
D = Falling action
E = Resolution

$\overline{AB}$ = We met for drinks

$\overline{AC}$ = And I was game

$\overline{AD}$ = Only took two minutes

$\overline{AE}$ = We don't talk

$\overline{BC}$ = We went to his place

$\overline{BD}$ = Like a metronome

$\overline{BE}$ = At least he had condoms

$\overline{CD}$ = I called a cab

$\overline{CE}$ = He showered

$\overline{DE}$ = And I snuck out

***Figure 22*: Carbon-based life forms**

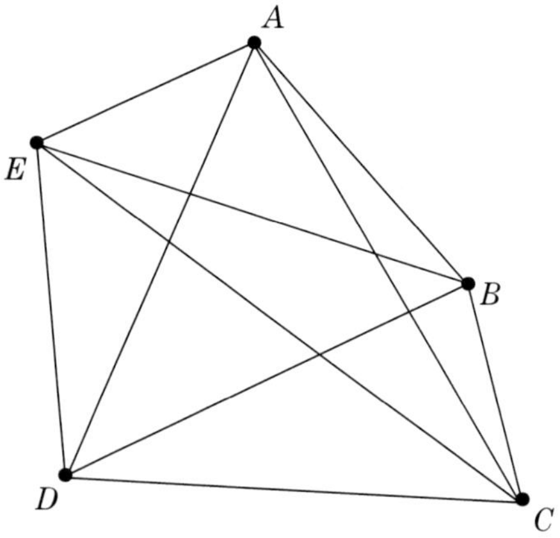

A = Mammalian
B = Avian
C = Fungal
D = Viral
E = Reptilian

$\overline{AB}$ = Easter bunny

$\overline{AC}$ = Athlete's foot

$\overline{AD}$ = Common cold

$\overline{AE}$ = Your evil ex

$\overline{BC}$ = Histoplasmosis

$\overline{BD}$ = Bird flu

$\overline{BE}$ = Dinosaurs

$\overline{CD}$ = No antibiotics for you

$\overline{CE}$ = Turtles in captivity

$\overline{DE}$ = Online trolls

Figure 23: Cyborg attachments

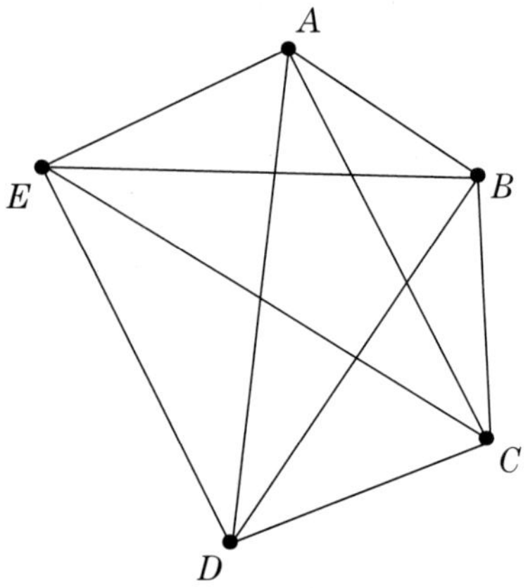

A = Orthodontia
B = Intrauterine devices
C = Pacemakers
D = Pedometers
E = Bluetooth headsets

$\overline{AB}$ = Traumatic puberty

$\overline{AC}$ = Trouble at airport security

$\overline{AD}$ = Weight loss

$\overline{AE}$ = Nerds

$\overline{BC}$ = May to December pairings

$\overline{BD}$ = Taking protective steps

$\overline{BE}$ = Corner offices

$\overline{CD}$ = No second heart attack

$\overline{CE}$ = Inside voices

$\overline{DE}$ = Mall walkers

Figure 24: The magic of familiars

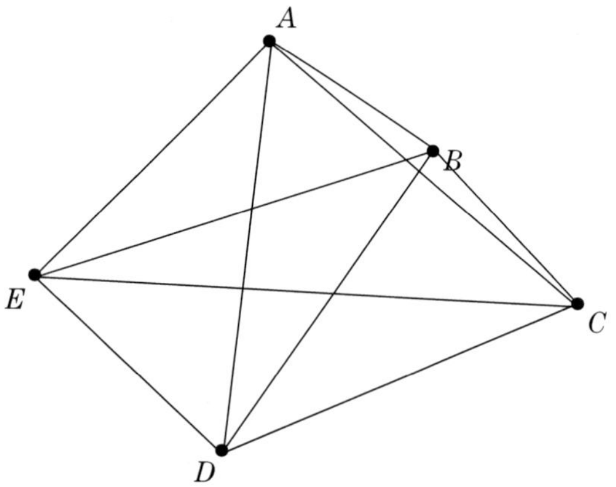

A = Cats
B = Dogs
C = Hamsters
D = Birds
E = Horses

$\overline{AB}$ = Downpour

$\overline{AC}$ = Poop in a box

$\overline{AD}$ = Circle of life

$\overline{AE}$ = Working barn

$\overline{BC}$ = Happy to run in circles

$\overline{BD}$ = Retrievers

$\overline{BE}$ = Doomed fox

$\overline{CD}$ = Shoebox coffins

$\overline{CE}$ = Mostly bad bets

$\overline{DE}$ = Pegasus

***Figure 25*: Choose your weapons wisely**

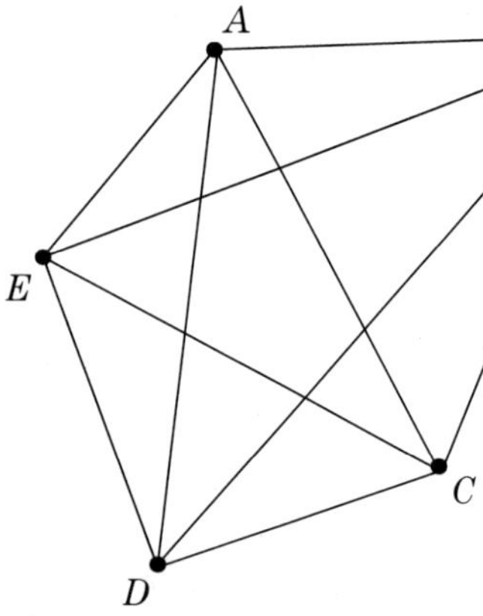

A = Words
B = Knives
C = Sex
D = Guns
E = Money

$\overline{AB}$ = Insults

$\overline{AC}$ = Erotica

$\overline{AD}$ = The Worst Amendment

$\overline{AE}$ = Contracts

$\overline{BC}$ = Fear play

$\overline{BD}$ = Insecurity

$\overline{BE}$ = Muggings

$\overline{CD}$ = Movies for teenage boys

$\overline{CE}$ = Capitalism

$\overline{DE}$ = Gangsters

***Figure 26*: Game time**

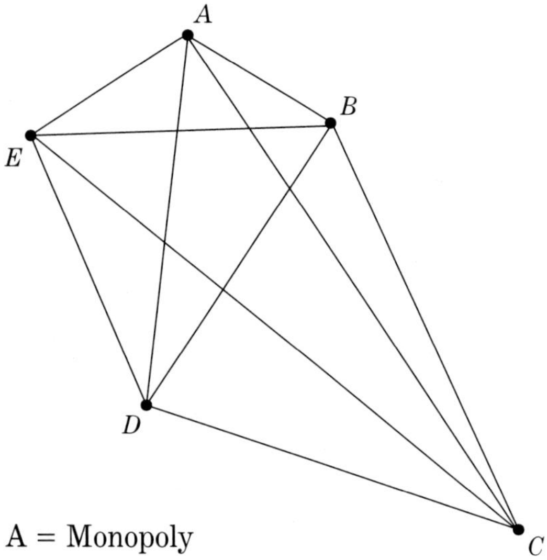

A = Monopoly
B = Sorry
C = Candyland
D = Life
E = Pandemic

$\overline{AB}$ = Comcast

$\overline{AC}$ = Locked cookie jar

$\overline{AD}$ = Our rare planet

$\overline{AE}$ = Pharmaceutical patents

$\overline{BC}$ = Nut allergies

$\overline{BD}$ = Deathbed regrets

$\overline{BE}$ = NIMBYs

$\overline{CD}$ = Rich parents

$\overline{CE}$ = Metabolic syndrome

$\overline{DE}$ = Invasive species

Figure 27: Payments adults make

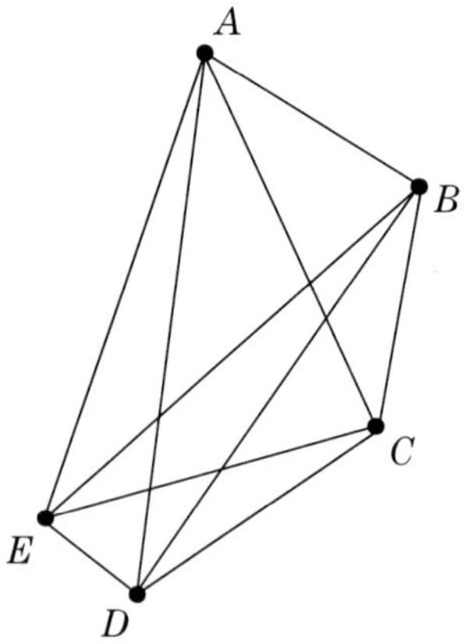

A = Insurance
B = Rent
C = Alimony
D = Storage unit
E = Dues

$\overline{AB}$ = In case of burglars

$\overline{AC}$ = Prenuptial agreement

$\overline{AD}$ = Copies of all my papers

$\overline{AE}$ = So my heirs have something

$\overline{BC}$ = My own quiet home

$\overline{BD}$ = The cost of safe secrets

$\overline{BE}$ = Community pool access

$\overline{CD}$ = My space is worth it

$\overline{CE}$ = Lessons learned

$\overline{DE}$ = A room of one's own

Figure 28: Mixing artistic media

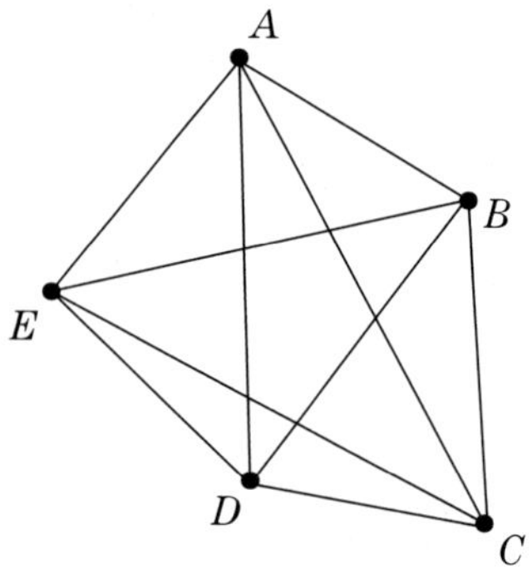

A = Marble
B = Oil
C = Music
D = Language
E = Make-up

$\overline{AB}$ = Kobe beef

$\overline{AC}$ = The bust on the piano

$\overline{AD}$ = Headstone typography

$\overline{AE}$ = Airbrushed abs

$\overline{BC}$ = Rosin

$\overline{BD}$ = Medieval French

$\overline{BE}$ = Lip gloss

$\overline{CD}$ = Lyrics

$\overline{CE}$ = Stagecraft

$\overline{DE}$ = Wordless flirtation

Figure 29: The same back and forth

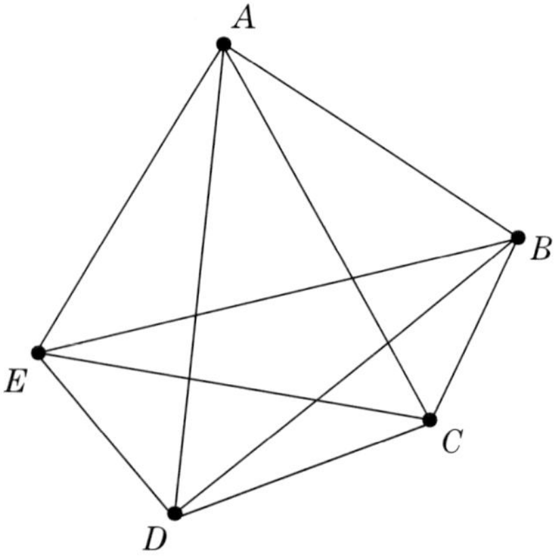

A = Race car
B = Kayak
C = Radar
D = Sagas
E = Wow

$\overline{AB}$ = Single scull

$\overline{AC}$ = Bird dog

$\overline{AD}$ = Around the world in 80 days

$\overline{AE}$ = Crashes

$\overline{BC}$ = GPS

$\overline{BD}$ = Viking adventures

$\overline{BE}$ = Whitewater

$\overline{CD}$ = Smokey & the Bandit

$\overline{CE}$ = Salt flats

$\overline{DE}$ = Icelandic magic

***Figure 30*: Nice work, breeders**

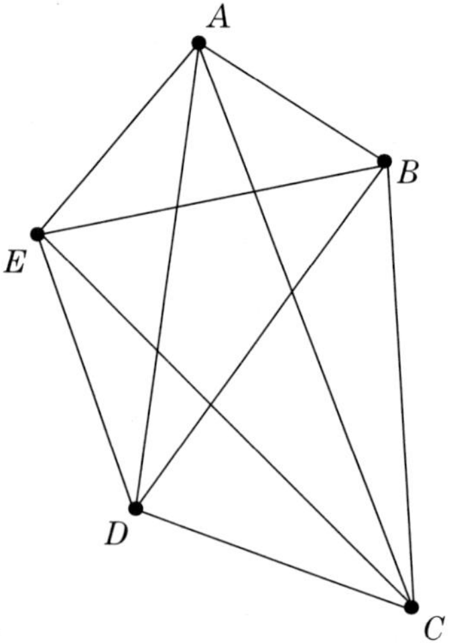

A = Retriever

B = Great Dane

C = Chihuahua

D = Poodle

E = Pug

$\overline{AB}$ = Good boy

$\overline{AC}$ = Good boy

$\overline{AD}$ = Good boy

$\overline{AE}$ = Good boy

$\overline{BC}$ = Good boy

$\overline{BD}$ = Good boy

$\overline{BE}$ = Good boy

$\overline{CD}$ = Good boy

$\overline{CE}$ = Good boy

$\overline{DE}$ = Good boy

***Figure 31*: Writing in stones**

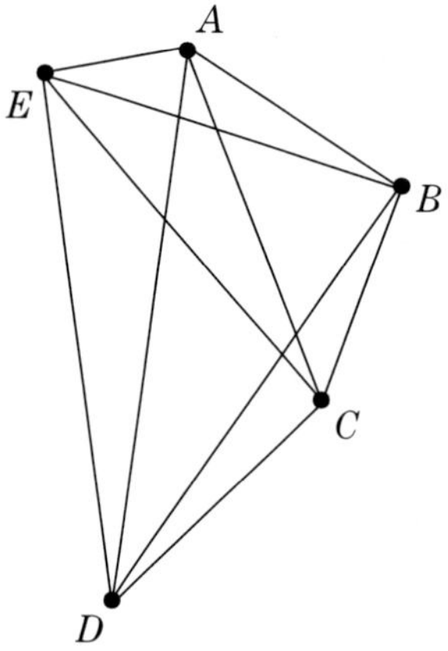

A = Metamorphic
B = Sedimentary
C = Igneous
D = Manufactured
E = Kidney

$\overline{AB}$ = Binge watching

$\overline{AC}$ = Overnight success

$\overline{AD}$ = Ad campaigns

$\overline{AE}$ = Anonymous donor

$\overline{BC}$ = Black sand beaches

$\overline{BD}$ = Fordite

$\overline{BE}$ = Calcium

$\overline{CD}$ = Baking soda & vinegar

$\overline{CE}$ = Sudden pain

$\overline{DE}$ = A scientist is trying

***Figure 32*: "It just went off."**

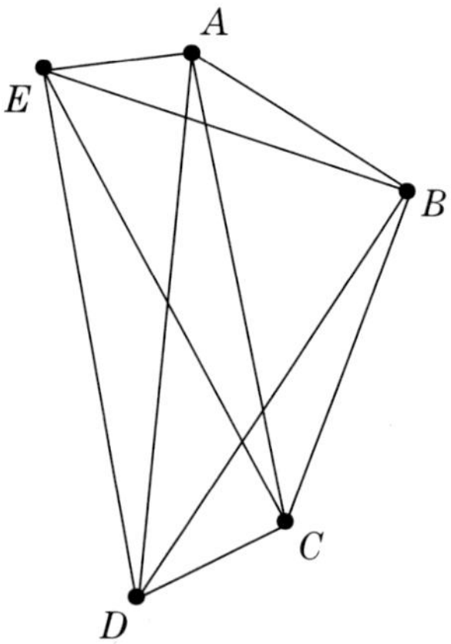

A = Neurons
B = Guns
C = Tempers
D = Volcanoes
E = Wanderers

$\overline{AB}$ = "I think I just shot him."

$\overline{AC}$ = "I think you should go."

$\overline{AD}$ = "I think we should run."

$\overline{AE}$ = "I don't think we're lost."

$\overline{BC}$ = "I wasn't thinking."

$\overline{BD}$ = "I heard a loud boom."

$\overline{BE}$ = "I don't hitchhike."

$\overline{CD}$ = "I fear mother nature."

$\overline{CE}$ = "I told you to bring a map."

$\overline{DE}$ = "Look, an island ahead."

***Figure 33*: Modern utilitarianism**

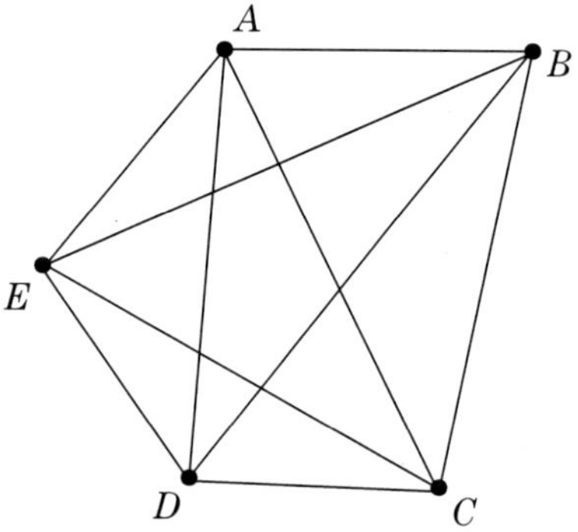

A = Light
B = Heat
C = Water
D = Waste
E = Internet

$\overline{AB}$ = Candles

$\overline{AC}$ = Hydroelectric plants

$\overline{AD}$ = Urban sky glow

$\overline{AE}$ = Soft blue

$\overline{BC}$ = Tea

$\overline{BD}$ = Car exhaust

$\overline{BE}$ = Manufactured outrage

$\overline{CD}$ = Flushing

$\overline{CE}$ = Streams

$\overline{DE}$ = Comment sections

***Figure 34*: What we raise**

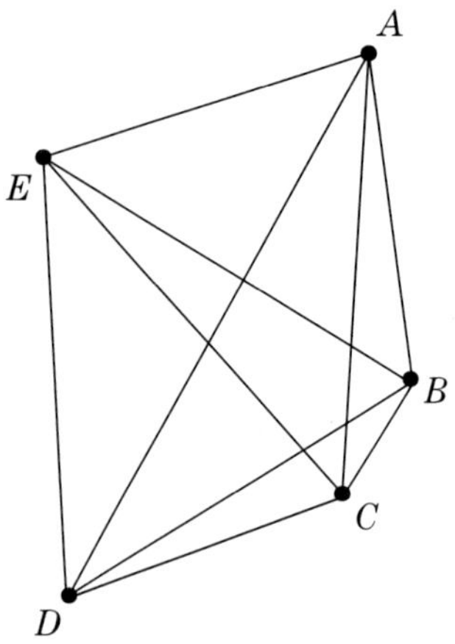

A = Children
B = Voices
C = Concerns
D = Crops
E = Eyebrows

$\overline{AB}$ = Honesty

$\overline{AC}$ = Futurists

$\overline{AD}$ = Exploited labor

$\overline{AE}$ = Too many or too few

$\overline{BC}$ = Assertiveness

$\overline{BD}$ = Agricultural lobbies

$\overline{BE}$ = Raconteurs

$\overline{CD}$ = Drought & pestilence

$\overline{CE}$ = Gossip

$\overline{DE}$ = Prize-winning pumpkins

Figure 35: Can go bad quickly

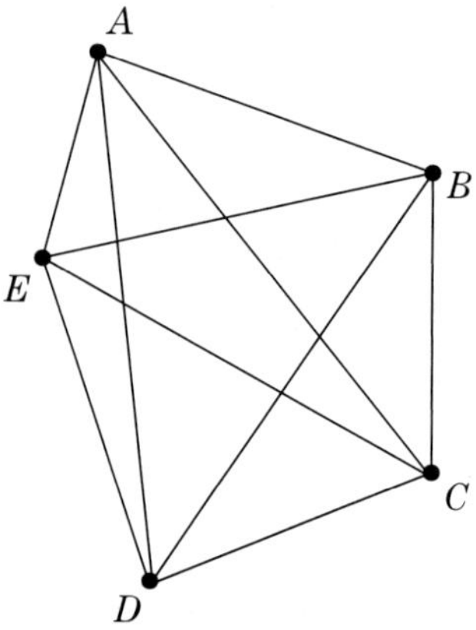

A = Milk

B = Eggs

C = Seeds

D = Hair

E = Moods

$\overline{AB}$ = Feminine

$\overline{AC}$ = On vegan cereal

$\overline{AD}$ = Conditioner

$\overline{AE}$ = Postpartum depression

$\overline{BC}$ = Potential

$\overline{BD}$ = Brunch sent back

$\overline{BE}$ = PMS

$\overline{CD}$ = Dandelions

$\overline{CE}$ = Gardening

$\overline{DE}$ = Drastic cuts

***Figure 36*: Will it get you high?**

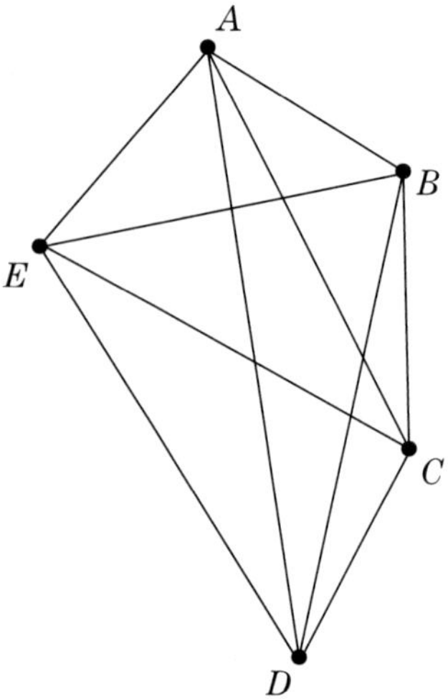

A = Acid
B = Booze
C = Weed
D = Pills
E = Religion

$\overline{AB}$ = Tart wine

$\overline{AC}$ = Dandelion salad

$\overline{AD}$ = Heartburn

$\overline{AE}$ = Kool-Aid

$\overline{BC}$ = Wormwood

$\overline{BD}$ = Suicide attempt

$\overline{BE}$ = Communion wine

$\overline{CD}$ = Alfalfa

$\overline{CE}$ = Incense

$\overline{DE}$ = Faith in modern medicine

***Figure 37*: Ancient elements**

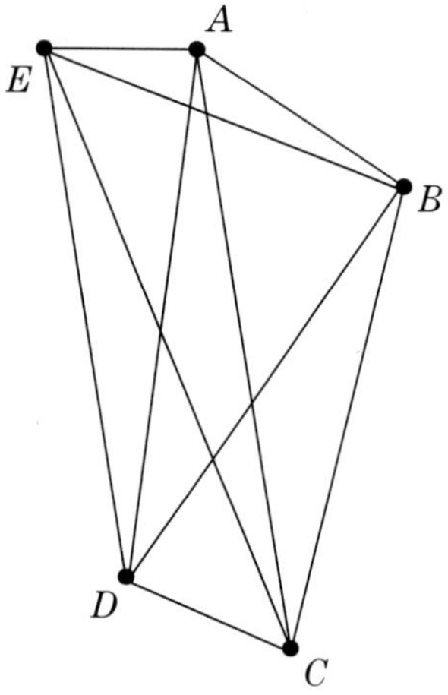

A = Fire
B = Earth
C = Metal
D = Water
E = Wood

$\overline{AB}$ = Lava

$\overline{AC}$ = Weld

$\overline{AD}$ = Alcohol

$\overline{AE}$ = Stove

$\overline{BC}$ = Ore

$\overline{BD}$ = Mud

$\overline{BE}$ = Root

$\overline{CD}$ = Pipe

$\overline{CE}$ = Flute

$\overline{DE}$ = Smoke

***Figure 38*: So sappy**

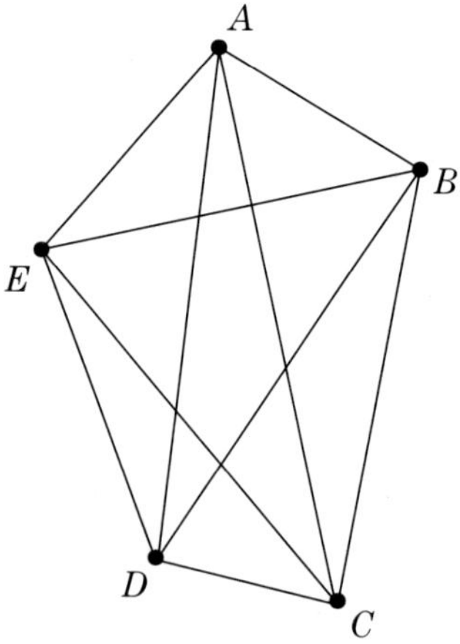

A = Sunset photos
B = Maple syrup
C = Love song lyrics
D = Soap operas
E = Valentines

$\overline{AB}$ = Vermont bed and breakfast

$\overline{AC}$ = Wedding invitations

$\overline{AD}$ = Closing credits

$\overline{AE}$ = Wish you were here

$\overline{BC}$ = Stuck on you

$\overline{BD}$ = Cozy mysteries

$\overline{BE}$ = Worst candy in the box

$\overline{CD}$ = Housewife catnip

$\overline{CE}$ = Cheap cards

$\overline{DE}$ = Drama

Figure 39: Midday processing

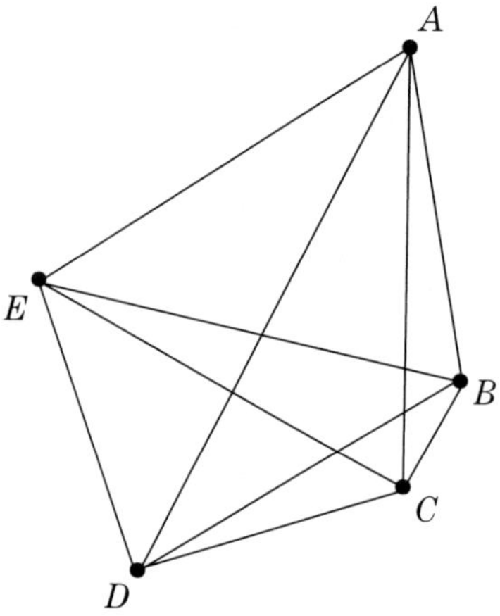

A = Paperwork
B = Feelings
C = Lunch
D = Experience
E = Data

$\overline{AB}$ = Love letters

$\overline{AC}$ = Dividing the tab

$\overline{AD}$ = Diplomas

$\overline{AE}$ = Receipts

$\overline{BC}$ = Gentle hunger

$\overline{BD}$ = Working nervous systems

$\overline{BE}$ = Sentient machines

$\overline{CD}$ = Fullness

$\overline{CE}$ = Calorie counts

$\overline{DE}$ = Sr. analysts

***Figure 40*: Footwork**

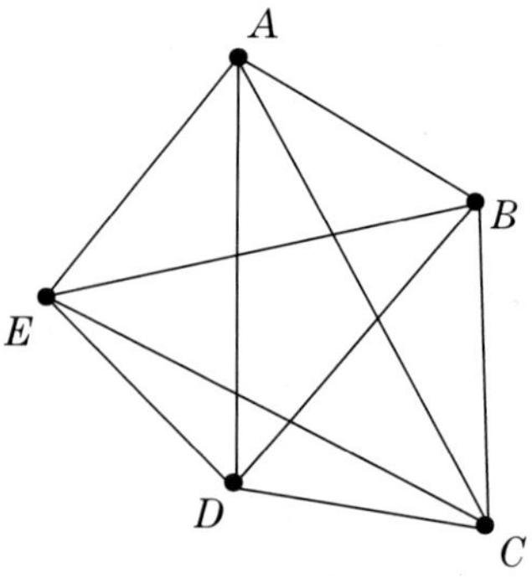

A = Sneaker
B = Loafer
C = Heel
D = Slipper
E = Flip-flop

$\overline{AB}$ = Truant

$\overline{AC}$ = Slacker

$\overline{AD}$ = Escapee

$\overline{AE}$ = Surprise party

$\overline{BC}$ = Parasite

$\overline{BD}$ = Went home early

$\overline{BE}$ = Burst of enthusiasm

$\overline{CD}$ = Thief

$\overline{CE}$ = Prodigal son

$\overline{DE}$ = Waterslide

***Figure 41*: Can I get you anything?**

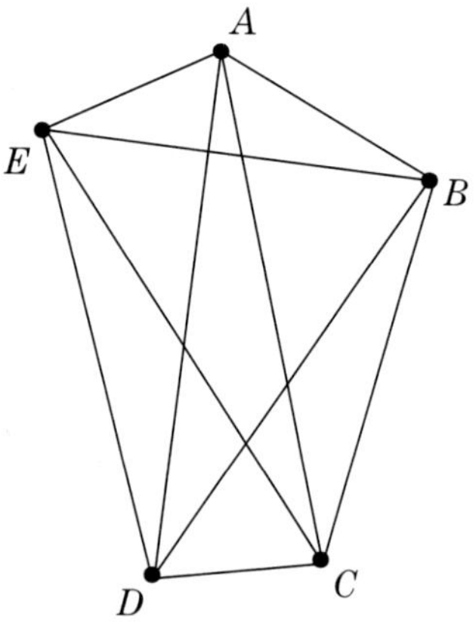

A = Coffee
B = Tea
C = Whisky
D = Milk
E = Juice

$\overline{AB}$ = Typical staff meeting

$\overline{AC}$ = Terrible staff meeting

$\overline{AD}$ = Latte

$\overline{AE}$ = Morning play date

$\overline{BC}$ = Toddy

$\overline{BD}$ = Brits

$\overline{BE}$ = Arnold Palmer

$\overline{CD}$ = Sneaky Pete

$\overline{CE}$ = Manhattan

$\overline{DE}$ = School lunch

***Figure 42*: Guns**

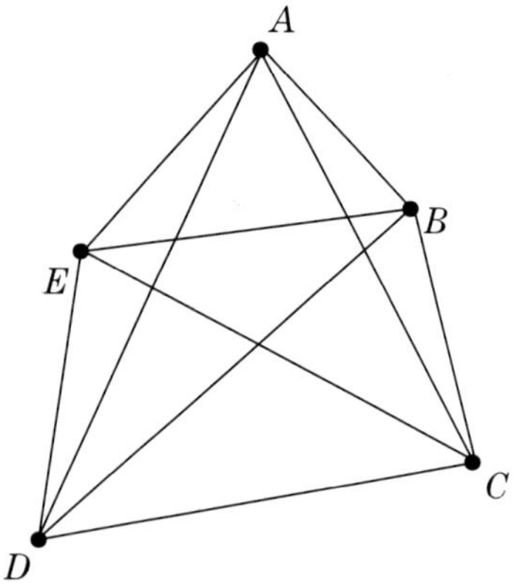

A = Glue
B = Water
C = Caulk
D = Smoking
E = Loaded

$\overline{AB}$ = Paper maché

$\overline{AC}$ = Sub-par renovation

$\overline{AD}$ = Desperate to get high

$\overline{AE}$ = Pistol at the horse stable

$\overline{BC}$ = Backsplash

$\overline{BD}$ = Bongs

$\overline{BE}$ = Beachfront property

$\overline{CD}$ = Bathtub candle mishap

$\overline{CE}$ = Messy guns

$\overline{DE}$ = Fine, fine cigars

***Figure 43*: What's eating you?**

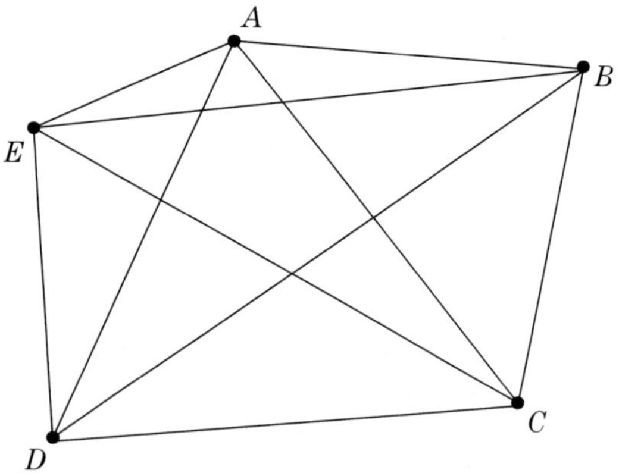

A = Tapeworms
B = Envy
C = Regret
D = Zombies
E = Time

$\overline{AB}$ = Thinness

$\overline{AC}$ = That steak tartare you had

$\overline{AD}$ = Occupational hazard

$\overline{AE}$ = Anemia

$\overline{BC}$ = FOMO

$\overline{BD}$ = Neighbor's locked shelter

$\overline{BE}$ = Seething

$\overline{CD}$ = Getting bitten

$\overline{CE}$ = Perspective

$\overline{DE}$ = Disintegration

Figure 44: What gets put down

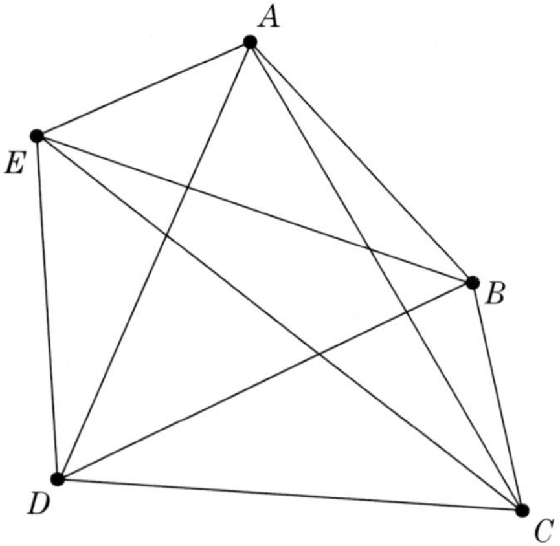

A = Rabid dogs
B = Brutal books
C = Tired babies
D = Car windows
E = The weapon

$\overline{AB}$ = Old Yeller

$\overline{AC}$ = Screaming mother

$\overline{AD}$ = Drive! Drive!

$\overline{AE}$ = Shotgun

$\overline{BC}$ = Dr. Seuss (again)

$\overline{BD}$ = On the Road

$\overline{BE}$ = Censorship

$\overline{CD}$ = Asleep in the car seat

$\overline{CE}$ = Lullaby

$\overline{DE}$ = Hammer

***Figure 45*: Flavors separated from taste**

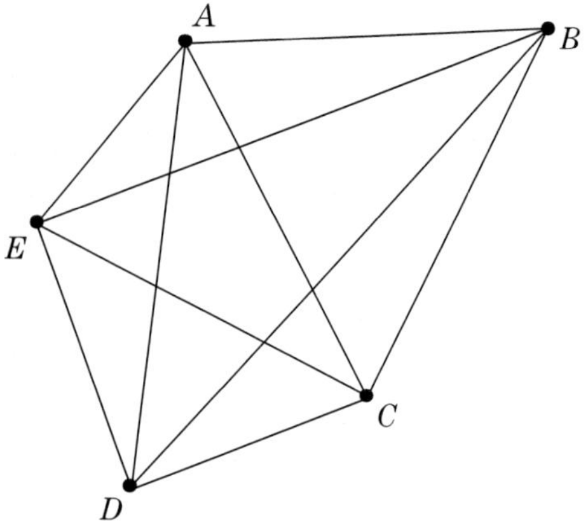

A = Salt

B = Pepper

C = Halloween

D = Cinnamon

E = Mint

$\overline{AB}$ = Pass

$\overline{AC}$ = Aging redheads

$\overline{AD}$ = Weak prank

$\overline{AE}$ = Artisanal caramels

$\overline{BC}$ = Strippers

$\overline{BD}$ = Chutzpah

$\overline{BE}$ = Gum

$\overline{CD}$ = Perfect French toast

$\overline{CE}$ = Masks halitosis

$\overline{DE}$ = Mojitos

Figure 46: How to get rid of pesky truths

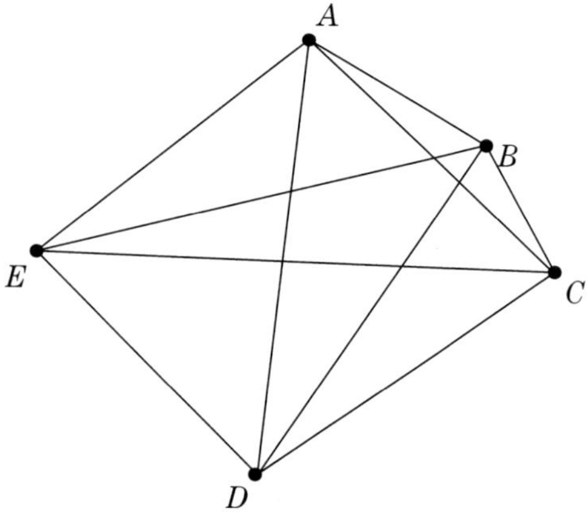

A = Bury
B = Ignore
C = Incinerate
D = Neglect
E = Ban

$\overline{AB}$ = Mass graves

$\overline{AC}$ = Tire fires

$\overline{AD}$ = Weedy graveyard

$\overline{AE}$ = Dangerous ideas

$\overline{BC}$ = Climate change denial

$\overline{BD}$ = Feral children

$\overline{BE}$ = Reading it anyway

$\overline{CD}$ = Poor campfire hygiene

$\overline{CE}$ = Must-reads

$\overline{DE}$ = Criminalizing poverty

Figure 47: Interpersonal transgressions

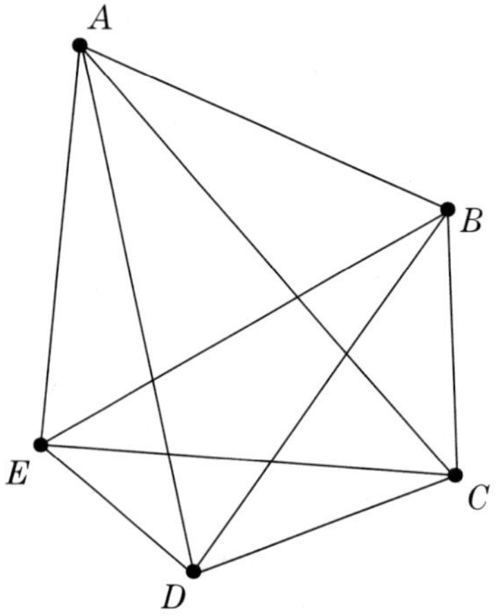

A = Insubordination
B = Stealing
C = Lying
D = Murder
E = Coveting

$\overline{AB}$ = Slacking at work

$\overline{AC}$ = Sticking it to The Man

$\overline{AD}$ = Patricide

$\overline{AE}$ = Striving

$\overline{BC}$ = Identity theft

$\overline{BD}$ = Motivated by life insurance

$\overline{BE}$ = Jewels gone missing

$\overline{CD}$ = Faking your own death

$\overline{CE}$ = “She’s not my type.”

$\overline{DE}$ = Cain & Abel

***Figure 48*: Senselessness**

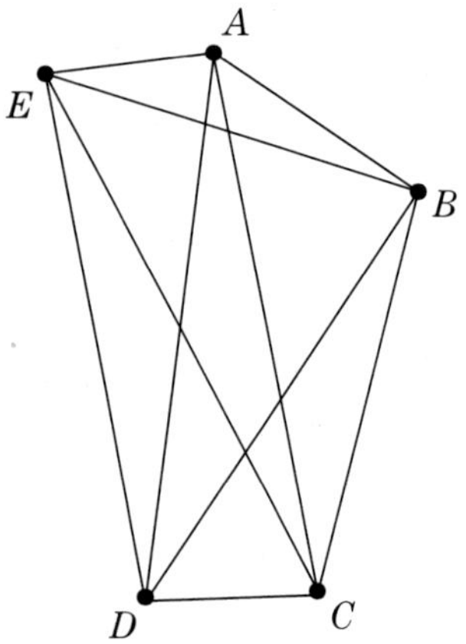

A = Deaf
B = Dumb
C = Blind
D = Numb
E = Tasteless

$\overline{AB}$ = Ideal secret-keeper

$\overline{AC}$ = Keller

$\overline{AD}$ = Shellshock

$\overline{AE}$ = The tone of an unfunny joke

$\overline{BC}$ = Reckless

$\overline{BD}$ = Indica

$\overline{BE}$ = Blockbuster movie

$\overline{CD}$ = Denial

$\overline{CE}$ = Racist grandpas

$\overline{DE}$ = Dental surgery

Figure 49: Physical pains

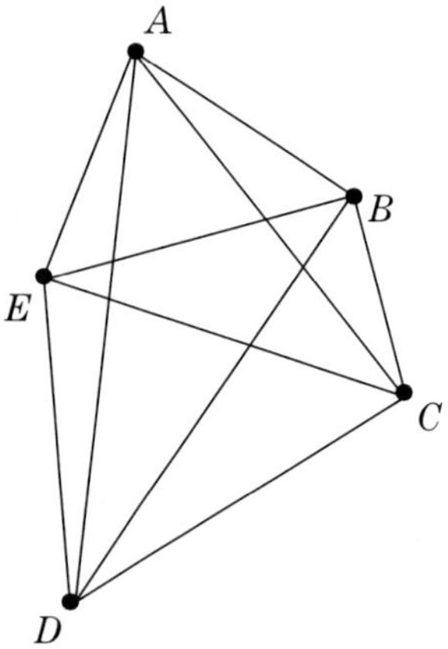

A = Burns
B = Scrapes
C = Fractures
D = Infections
E = Cuts

$\overline{AB}$ = Scabs

$\overline{AC}$ = Friendship-ending insults

$\overline{AD}$ = Bumpy scars

$\overline{AE}$ = Cauterized

$\overline{BC}$ = Falling off a log

$\overline{BD}$ = Road rash

$\overline{BE}$ = Misuse of cheese grater

$\overline{CD}$ = Sepsis

$\overline{CE}$ = Open-heart surgery

$\overline{DE}$ = Not enough soap and water

***Figure 50*: Probing ills**

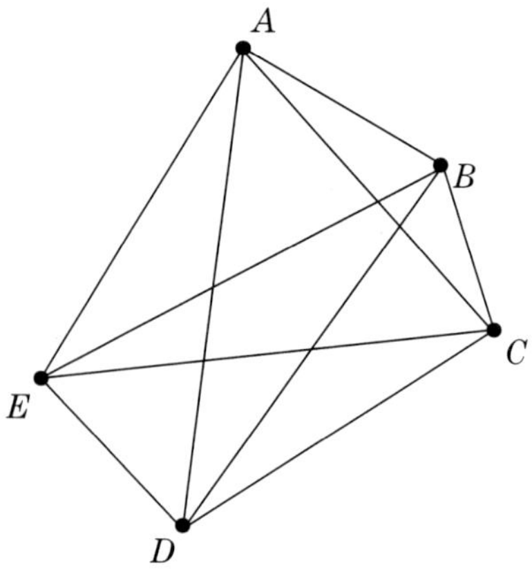

A = Mental
B = Physical
C = Moral
D = Environmental
E = Social

$\overline{AB}$ = Headaches

$\overline{AC}$ = Rationalizing

$\overline{AD}$ = Lead in water

$\overline{AE}$ = Emotional intelligence

$\overline{BC}$ = Body shaming

$\overline{BD}$ = Rock climbing

$\overline{BE}$ = Dancing

$\overline{CD}$ = Tree planting

$\overline{CE}$ = Sharing

$\overline{DE}$ = Parks

Figure 51: Senses at work

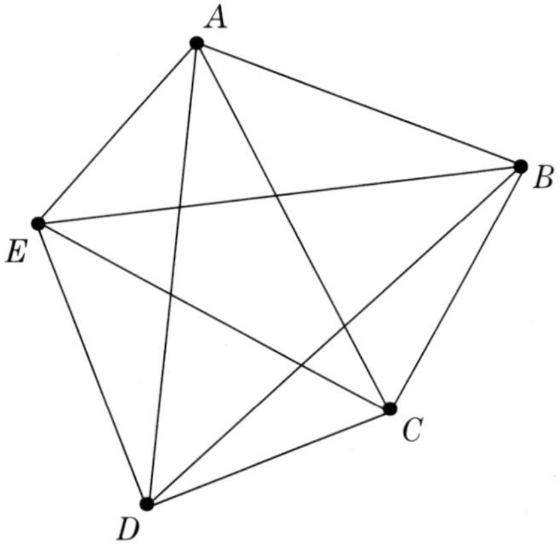

A = Sight
B = Sound
C = Taste
D = Touch
E = Smell

$\overline{AB}$ = Television

$\overline{AC}$ = Fine tailoring

$\overline{AD}$ = Contact lens insertion

$\overline{AE}$ = Salivation

$\overline{BC}$ = Music nerds

$\overline{BD}$ = Ridges on records

$\overline{BE}$ = Occupied bathroom stalls

$\overline{CD}$ = Chewing

$\overline{CE}$ = Subtle perfumes

$\overline{DE}$ = Embrace

***Figure 52*: Modern addictions**

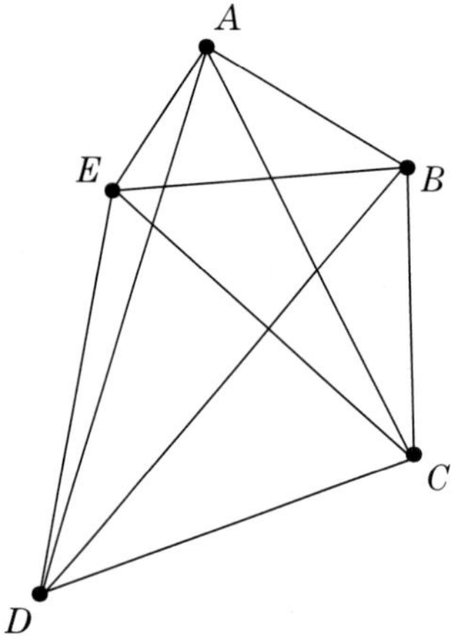

A = Carbohydrates
B = Television
C = Shopping
D = Alcohol
E = Sex

$\overline{AB}$ = Popcorn at the movies

$\overline{AC}$ = Food courts

$\overline{AD}$ = Beer

$\overline{AE}$ = Heart-shaped chocolates

$\overline{BC}$ = Commercials that work

$\overline{BD}$ = Lonely Fridays

$\overline{BE}$ = Pay-per-view

$\overline{CD}$ = Credit card debt

$\overline{CE}$ = Online dating

$\overline{DE}$ = Singles bars

Figure 53: Well, that sucks

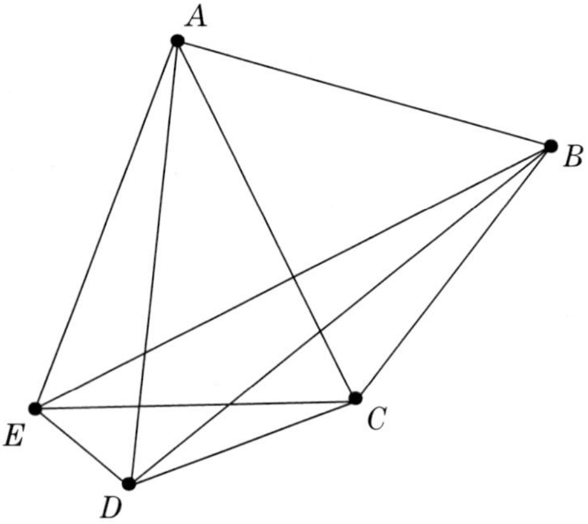

A = Rejection
B = Vampires
C = Death
D = Mediocrity
E = Leeches

$\overline{AB}$ = Bad blood

$\overline{AC}$ = The organ didn't take

$\overline{AD}$ = Failure to make the team

$\overline{AE}$ = Refusing to work for free

$\overline{BC}$ = Wooden stakes

$\overline{BD}$ = Dull fangs

$\overline{BE}$ = Bloodsucking

$\overline{CD}$ = Very short obituaries

$\overline{CE}$ = Medieval medicine

$\overline{DE}$ = Nepotism

Figure 54: Earning it takes time

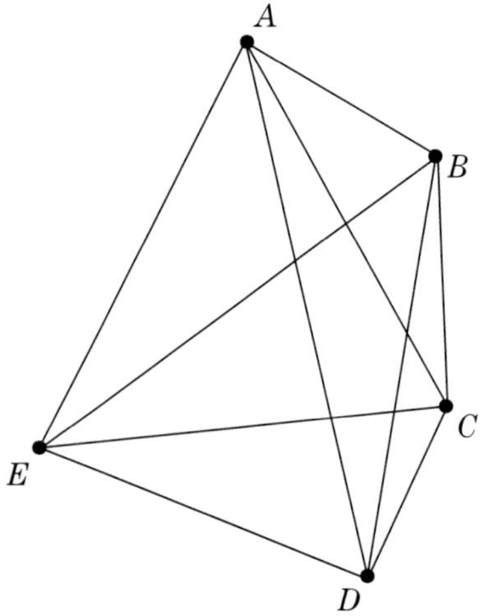

A = Money
B = Degrees
C = Trust
D = Respect
E = Time off

$\overline{AB}$ = Loans

$\overline{AC}$ = Credit

$\overline{AD}$ = America

$\overline{AE}$ = Leisure class

$\overline{BC}$ = Expertise

$\overline{BD}$ = Academia

$\overline{BE}$ = Sabbatical

$\overline{CD}$ = Admiration

$\overline{CE}$ = No curfew

$\overline{DE}$ = Mental health days

Figure 55: Full of it

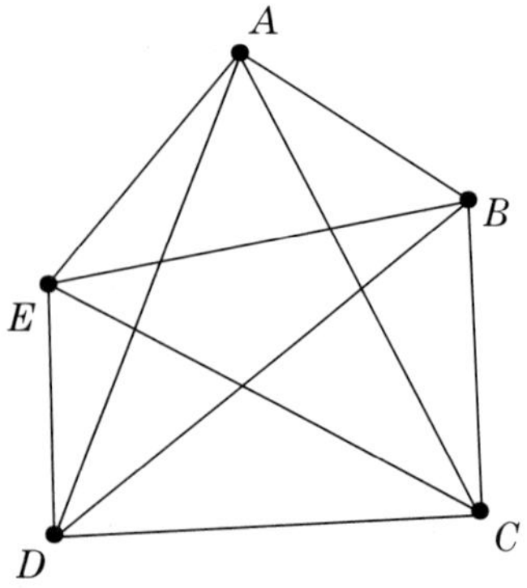

A = Grace
B = Potential
C = Desire
D = Shit
E = Wonder

$\overline{AB}$ = Hope

$\overline{AC}$ = Diligence

$\overline{AD}$ = Composure

$\overline{AE}$ = Appreciation

$\overline{BC}$ = Striving

$\overline{BD}$ = Fertilizer

$\overline{BE}$ = Curiosity

$\overline{CD}$ = Kink

$\overline{CE}$ = Wanderlust

$\overline{DE}$ = Witnessing

Figure 56: So you want to make art?

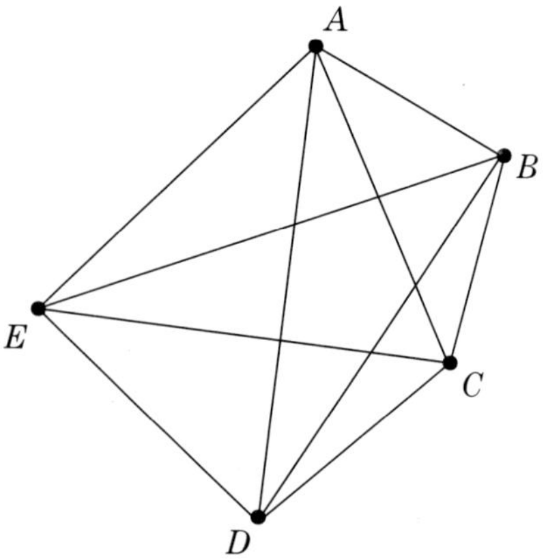

A = Annoyance
B = Surprise
C = Delight
D = Disgust
E = Disappointment

$\overline{AB}$ = He stole my idea.

$\overline{AC}$ = Someone stole his idea.

$\overline{AD}$ = He sold my idea.

$\overline{AE}$ = They say I stole my own idea.

$\overline{BC}$ = Someone likes my idea.

$\overline{BD}$ = My idea is mutilated.

$\overline{BE}$ = My idea is panned.

$\overline{CD}$ = My idea offends someone.

$\overline{CE}$ = His idea is panned.

$\overline{DE}$ = My idea is utterly ignored.

***Figure 57*: Old-school virtues**

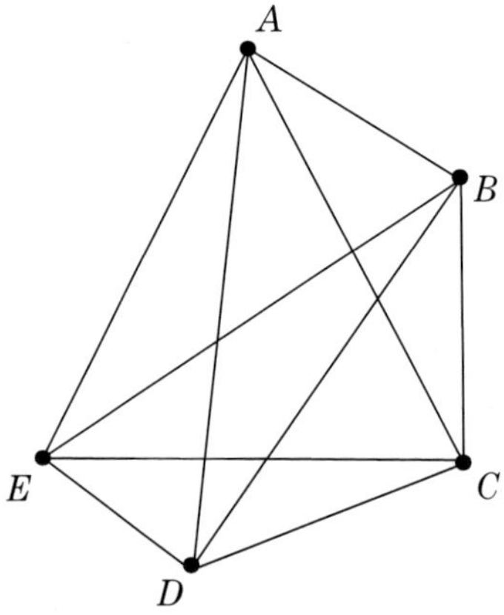

A = Friendship
B = Generosity
C = Chastity
D = Courtesy
E = Piety

$\overline{AB}$ = Forgiveness

$\overline{AC}$ = Boundaries

$\overline{AD}$ = Coworkers

$\overline{AE}$ = Church picnics

$\overline{BC}$ = Doubting the rumors

$\overline{BD}$ = After you

$\overline{BE}$ = Tithing

$\overline{CD}$ = Not staring

$\overline{CE}$ = Clergy vows

$\overline{DE}$ = Freedom of religion

***Figure 58*: Those are the breaks**

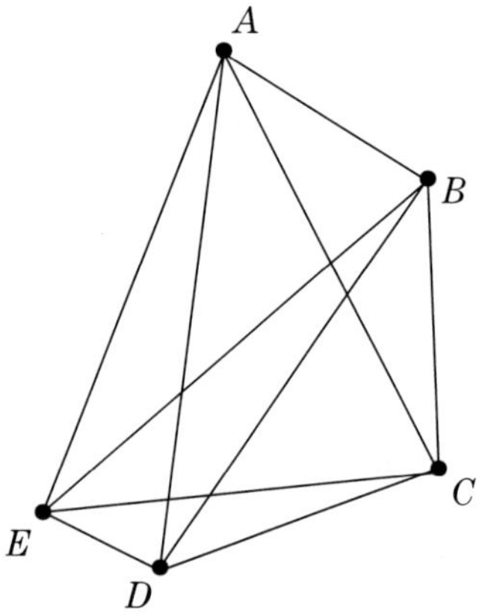

A = Heart
B = Jail
C = Coffee
D = Psychotic
E = Smoke

$\overline{AB}$ = Pressing charges

$\overline{AC}$ = Waking up

$\overline{AD}$ = Dropping charges

$\overline{AE}$ = Weakness

$\overline{BC}$ = Good cop interrogation

$\overline{BD}$ = Off his meds

$\overline{BE}$ = In the yard

$\overline{CD}$ = Whisky added

$\overline{CE}$ = Morning routine

$\overline{DE}$ = Arson

Figure 59: Are you allergic to that?

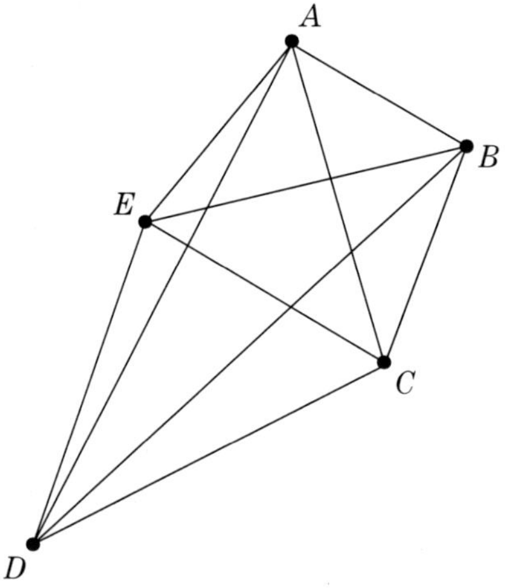

A = Bullshit
B = Nuts
C = Pollen
D = Work
E = Bees

$\overline{AB}$ = Conspiracy theorists

$\overline{AC}$ = Watery eyes

$\overline{AD}$ = Paper-pushing

$\overline{AE}$ = Colony collapse

$\overline{BC}$ = Masculinity

$\overline{BD}$ = Workaholics

$\overline{BE}$ = Nougat

$\overline{CD}$ = E

$\overline{CE}$ = Honey

$\overline{DE}$ = Females

Figure 60: New Year's resolutions

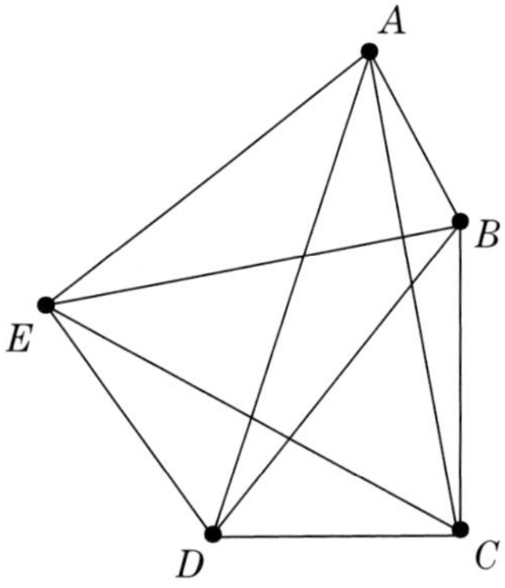

A = Thinner
B = Quieter
C = Wealthier
D = Less ambitious
E = More contented

$\overline{AB}$ = Demure

$\overline{AC}$ = Classy

$\overline{AD}$ = Decorative

$\overline{AE}$ = Harmless

$\overline{BC}$ = Judgmental

$\overline{BD}$ = Ordinary

$\overline{BE}$ = Smug

$\overline{CD}$ = Bored

$\overline{CE}$ = Philanthropic

$\overline{DE}$ = Unmotivated

***Figure 61*: What your tongue noticed**

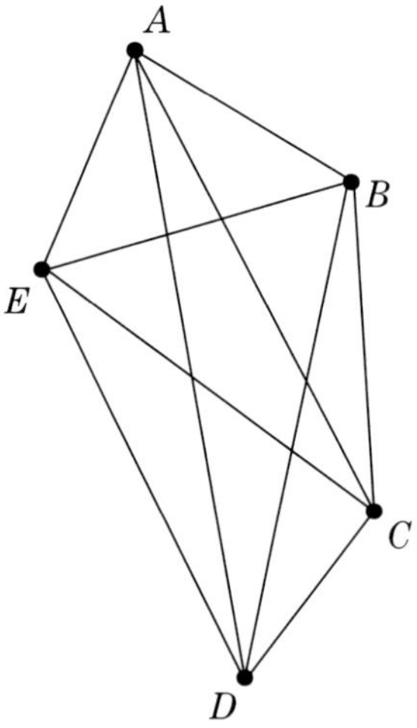

A = Sweet
B = Salty
C = Savory
D = Bitter
E = Sour

$\overline{AB}$ = PB&J

$\overline{AC}$ = Pies

$\overline{AD}$ = Raw rhubarb

$\overline{AE}$ = Chicken

$\overline{BC}$ = Honey mustard

$\overline{BD}$ = Unsuccessful pirates

$\overline{BE}$ = Buttermilk

$\overline{CD}$ = The best revenge

$\overline{CE}$ = Aged cheeses

$\overline{DE}$ = Mean spinster aunts

***Figure 62*: Saturday**

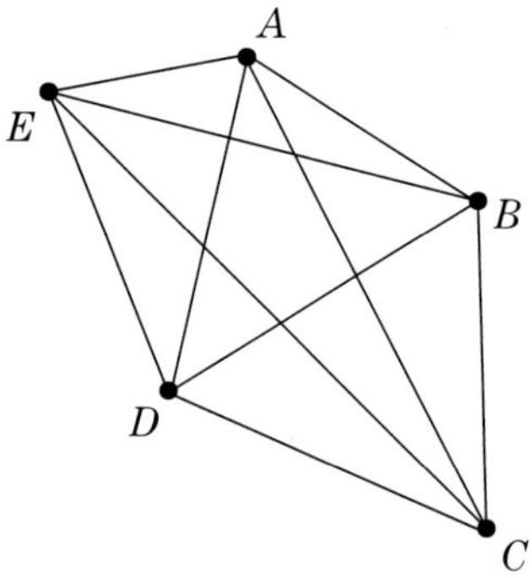

A = Eat
B = Drink
C = Dance
D = Kiss
E = Relax

$\overline{AB}$ = Milkshakes

$\overline{AC}$ = Wedding receptions

$\overline{AD}$ = Digesting lipstick

$\overline{AE}$ = Take-out

$\overline{BC}$ = Club

$\overline{BD}$ = Slurred affections

$\overline{BE}$ = Nightcap

$\overline{CD}$ = Romance

$\overline{CE}$ = Nobody's watching

$\overline{DE}$ = It's reciprocated

Figure 63: Ghost hobbies

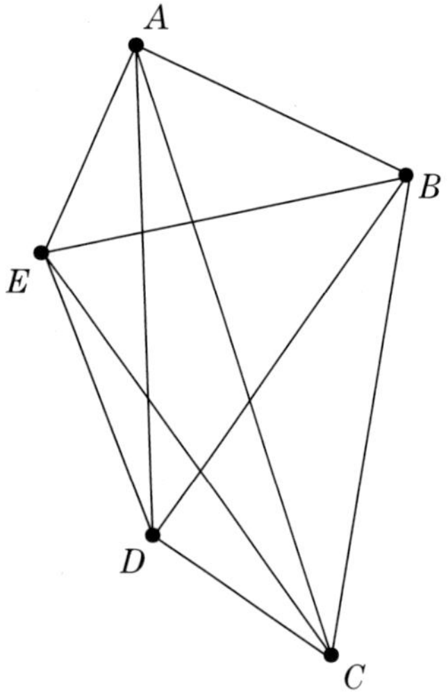

A = Peeping
B = Laughing
C = Rotting
D = Haunting
E = Avenging

$\overline{AB}$ = Police are called

$\overline{AC}$ = Sniffing old milk

$\overline{AD}$ = Naughty ghosts

$\overline{AE}$ = Unforgettable scenes

$\overline{BC}$ = Friendly zombie

$\overline{BD}$ = Banshee

$\overline{BE}$ = Last laugh

$\overline{CD}$ = The Tell-Tale Heart

$\overline{CE}$ = Food safety inspectors

$\overline{DE}$ = Witnessing the execution

***Figure 64*: What does a good day feel like?**

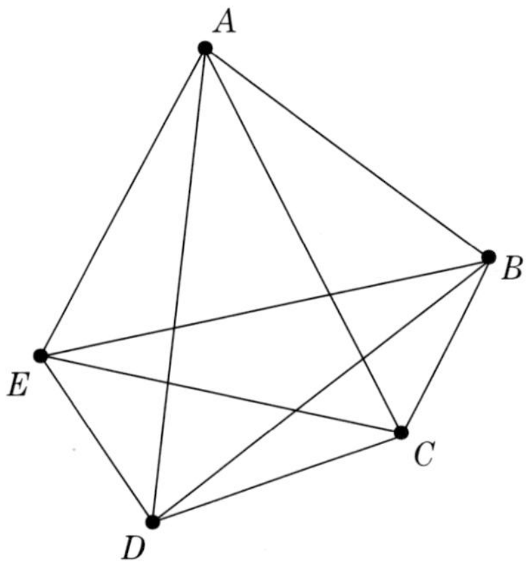

A = Fairness
B = Contentment
C = Compassion
D = Humility
E = Gratitude

$\overline{AB}$ = Good deal

$\overline{AC}$ = Justice

$\overline{AD}$ = Forgiveness

$\overline{AE}$ = Reciprocated efforts

$\overline{BC}$ = Snuggling

$\overline{BD}$ = Self-acceptance

$\overline{BE}$ = Equilibrium

$\overline{CD}$ = Mercy

$\overline{CE}$ = Hugging your loved ones

$\overline{DE}$ = Thank the Academy

***Figure 65*: Approaches to everything**

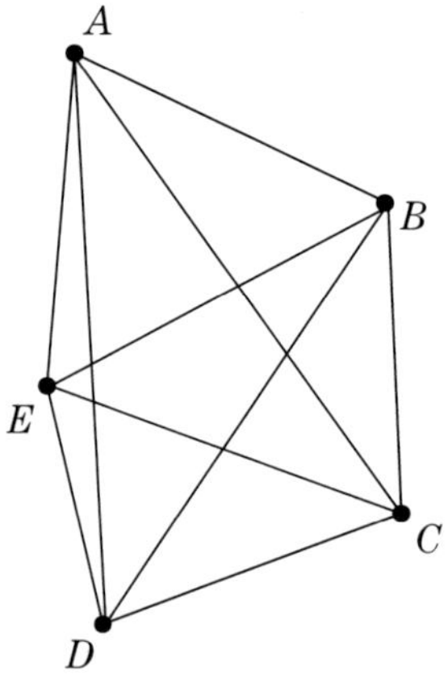

A = Fascination
B = Fear
C = Resignation
D = Hope
E = Ambivalence

$\overline{AB}$ = Serial killer documentaries

$\overline{AC}$ = Inspecting your wrinkles

$\overline{AD}$ = Scientific research

$\overline{AE}$ = Unrequited crush

$\overline{BC}$ = Staying at a rotten job

$\overline{BD}$ = Cautious optimism

$\overline{BE}$ = Nothing to lose

$\overline{CD}$ = Two weeks notice

$\overline{CE}$ = Ennui

$\overline{DE}$ = Coin toss

Figure 66: Vices

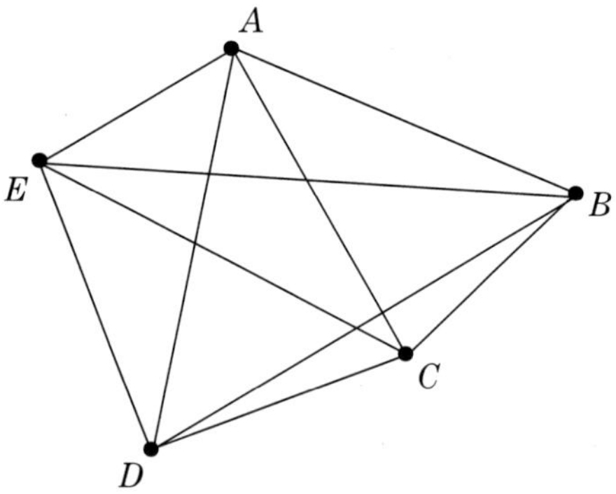

A = Lust
B = Rage
C = Greed
D = Attachment
E = Conceit

$\overline{AB}$ = Entitled celibates

$\overline{AC}$ = Harem

$\overline{AD}$ = Monogamy

$\overline{AE}$ = Narcissus

$\overline{BC}$ = Jealousy

$\overline{BD}$ = Slavery

$\overline{BE}$ = Ignored princes

$\overline{CD}$ = Misers

$\overline{CE}$ = Conspicuous consumption

$\overline{DE}$ = Smug marrieds

Figure 67: Prerequisites for school

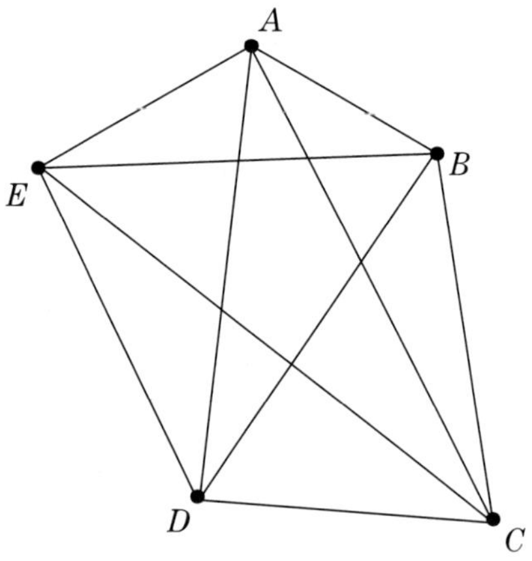

A = Food

B = Water

C = Shelter

D = Clothing

E = Love

$\overline{AB}$ = Soup

$\overline{AC}$ = Gingerbread house

$\overline{AD}$ = Edible underpants

$\overline{AE}$ = Dinner date

$\overline{BC}$ = Houseboat

$\overline{BD}$ = Bikini

$\overline{BE}$ = Amniotic fluid

$\overline{CD}$ = Overcoat

$\overline{CE}$ = Hospitality

$\overline{DE}$ = Hand-knit just for you

***Figure 68*: Things you need on your person**

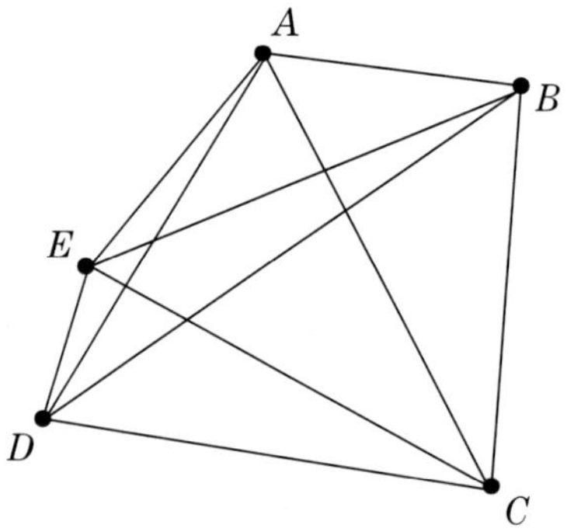

A = Time
B = Keys
C = Touch
D = Hope
E = Motivation

$\overline{AB}$ = Antique clocks

$\overline{AC}$ = Massage appointment

$\overline{AD}$ = Darkest hours

$\overline{AE}$ = Personal best marathon

$\overline{BC}$ = Pianos

$\overline{BD}$ = Prayers

$\overline{BE}$ = What you have to prove

$\overline{CD}$ = Reciprocated embrace

$\overline{CE}$ = Getting your buttons pushed

$\overline{DE}$ = Optimism

Figure 69: Fields of knowledge

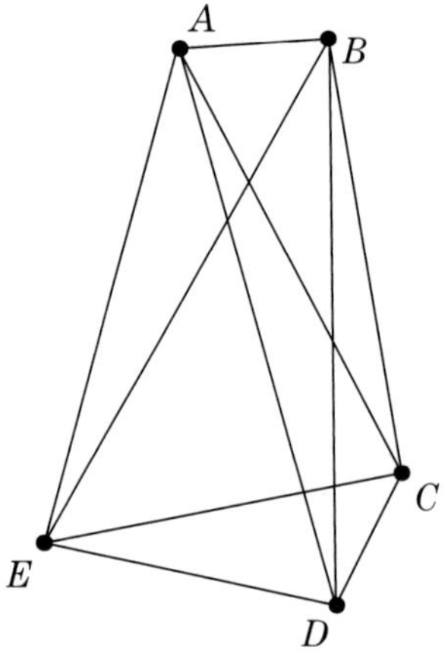

A = Sociology
B = Anatomy
C = History
D = Poetics
E = Gymnastics

$\overline{AB}$ = Fashion

$\overline{AC}$ = Racism

$\overline{AD}$ = Canon

$\overline{AE}$ = Etiquette

$\overline{BC}$ = Evolution

$\overline{BD}$ = Grammar

$\overline{BE}$ = Flexibility

$\overline{CD}$ = Linguistics

$\overline{CE}$ = Olympics

$\overline{DE}$ = Metaphor

***Figure 70*: Want to go to a party?**

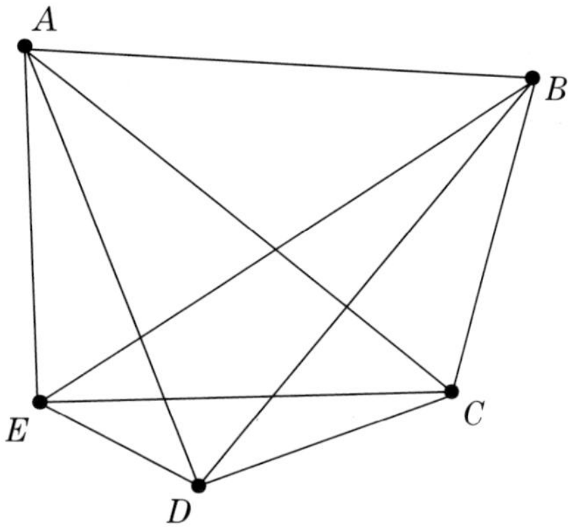

A = Birthday
B = Keg
C = Political
D = Ugly sweater
E = Office

$\overline{AB}$ = 21st

$\overline{AC}$ = July 4th

$\overline{AD}$ = December 25th

$\overline{AE}$ = Feigned enthusiasm

$\overline{BC}$ = Fundraiser

$\overline{BD}$ = Tailgating

$\overline{BE}$ = Law suit

$\overline{CD}$ = Convention schwag

$\overline{CE}$ = Inauguration

$\overline{DE}$ = Pseudo-fun

Figure 71: Mother Nature strikes back

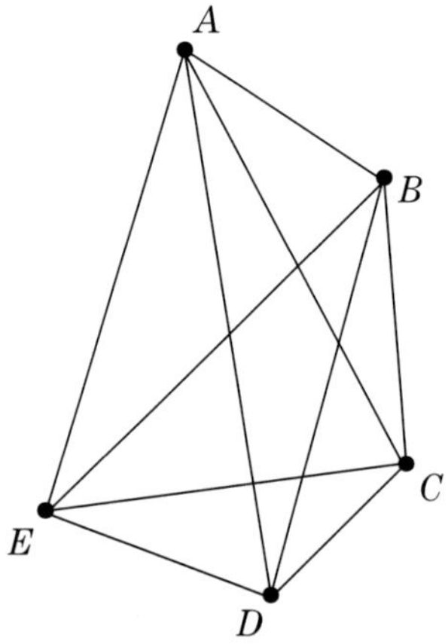

A = Fire
B = Flood
C = Pestilence
D = Earthquake
E = Eruption

$\overline{AB}$ = Minot, 2015

$\overline{AC}$ = London, 1666

$\overline{AD}$ = San Francisco, 1906

$\overline{AE}$ = Hiroshima, 1945

$\overline{BC}$ = Sumatra, 1833

$\overline{BD}$ = Santa Barbara, 1925

$\overline{BE}$ = Old Faithful, frequently

$\overline{CD}$ = Haiti, 2010

$\overline{CE}$ = Monrovia, 2014

$\overline{DE}$ = Mount Fuji, 1707

Figure 72: Wherever you go, there you are

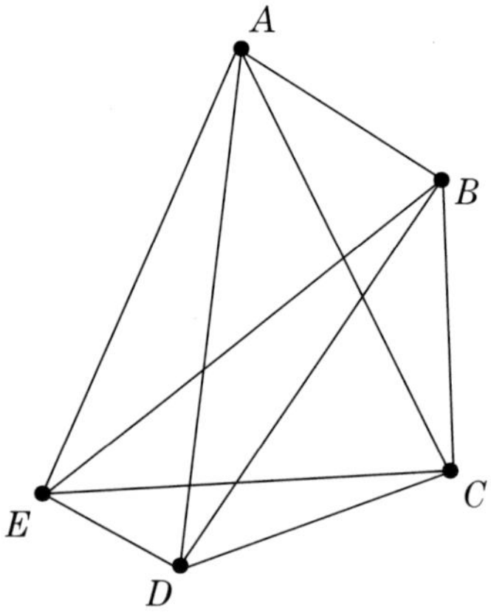

A = Utopia
B = Dystopia
C = Oblivion
D = Suburbia
E = Academia

$\overline{AB}$ = Drastically unequal societies

$\overline{AC}$ = Drugs

$\overline{AD}$ = Only in the brochure

$\overline{AE}$ = Sabbatical

$\overline{BC}$ = After the mushroom clouds

$\overline{BD}$ = Lifetime movies

$\overline{BE}$ = Loss of funding

$\overline{CD}$ = Big box stores

$\overline{CE}$ = Dropping out

$\overline{DE}$ = MRS degrees

Figure 73: Foundational matters

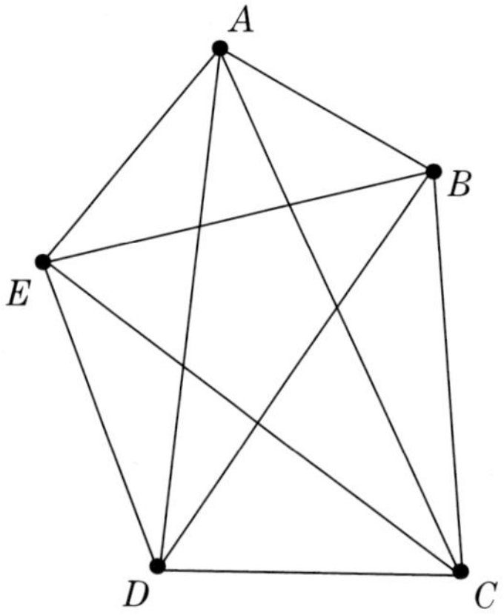

A = Solid
B = Liquid
C = Gas
D = Plasma
E = Spirit

$\overline{AB}$ = Non-homogenized milk

$\overline{AC}$ = Deposition

$\overline{AD}$ = Scab

$\overline{AE}$ = Confidence

$\overline{BC}$ = Diarrhea

$\overline{BD}$ = Donated for $20

$\overline{BE}$ = Cocktails

$\overline{CD}$ = Embolism

$\overline{CE}$ = Road trip

$\overline{DE}$ = Ectoplasm

***Figure 74*: Let's talk real estate**

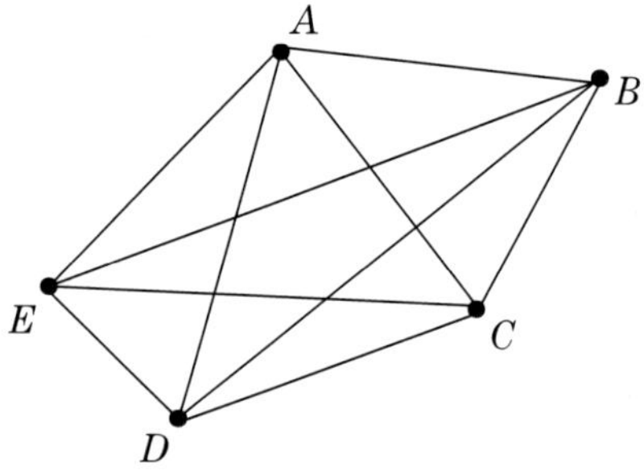

A = Location
B = Size
C = Functionality
D = Price
E = Availability

$\overline{\mathrm{AB}}$ = Smaller downtown

$\overline{\mathrm{AC}}$ = Airplane hangars

$\overline{\mathrm{AD}}$ = Too cheap means ghosts

$\overline{\mathrm{AE}}$ = Mr. & Mrs. Right Now

$\overline{\mathrm{BC}}$ = Know how to use it?

$\overline{\mathrm{BD}}$ = Per square foot

$\overline{\mathrm{BE}}$ = Niche online retailers

$\overline{\mathrm{CD}}$ = Beware the 5 bed, half-bath

$\overline{\mathrm{CE}}$ = Fixer-uppers

$\overline{\mathrm{DE}}$ = Cheap goes fast

Figure 75: Economics in play

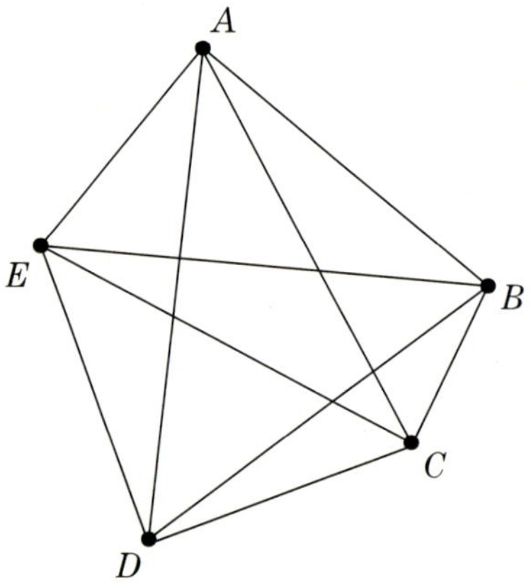

A = Mercantile
B = Industrial
C = Knowledge
D = Gig
E = Post-apocalyptic

$\overline{AB}$ = Widgets

$\overline{AC}$ = Webinars

$\overline{AD}$ = Garage sales

$\overline{AE}$ = Human organs

$\overline{BC}$ = Operating manuals

$\overline{BD}$ = Ditch digging

$\overline{BE}$ = Bunker welding

$\overline{CD}$ = Graphic design

$\overline{CE}$ = Maps of fresh water sources

$\overline{DE}$ = Pretty much

***Figure 76*: "Why are you giving me grief?"**

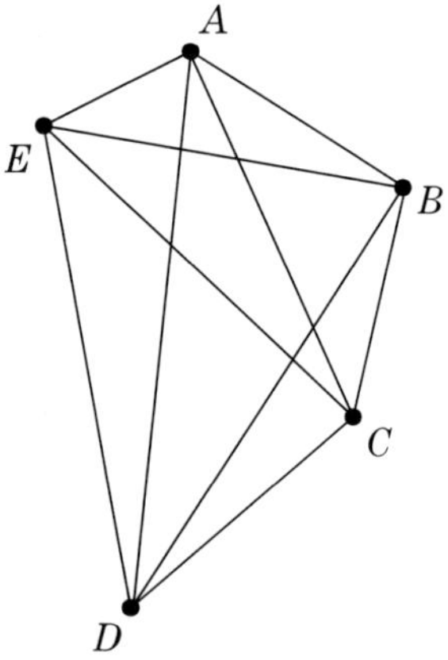

A = Denial
B = Anger
C = Bargaining
D = Depression
E = Acceptance

$\overline{AB}$ = "Liar."

$\overline{AC}$ = "I'm not paying that."

$\overline{AD}$ = "I'm fine."

$\overline{AE}$ = "That's against God's will."

$\overline{BC}$ = "You're trying to rip me off."

$\overline{BD}$ = "Those assholes don't care."

$\overline{BE}$ = "I don't have to like it."

$\overline{CD}$ = "Maybe I'll try tomorrow."

$\overline{CE}$ = "You have a deal."

$\overline{DE}$ = "I need to see a doctor."

Figure 77: Just fell into it

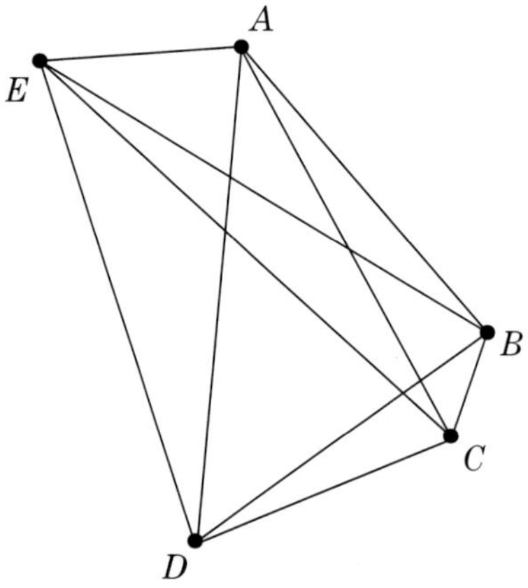

A = Debt
B = Despair
C = Temptation
D = Routine
E = Love

$\overline{AB}$ = Foreclosure

$\overline{AC}$ = Baubles

$\overline{AD}$ = Monthly payments

$\overline{AE}$ = Sending the kids to college

$\overline{BC}$ = Addiction

$\overline{BD}$ = Burnout

$\overline{BE}$ = Loss

$\overline{CD}$ = Commuting past a bakery

$\overline{CE}$ = Difficulty with fidelity

$\overline{DE}$ = Kiss before commuting

Figure 78: Water in other words

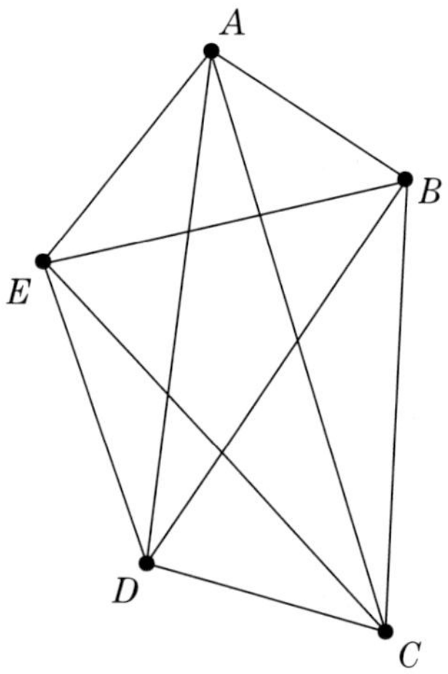

A = River
B = Lake
C = Ocean
D = Glacier
E = Cloud

$\overline{AB}$ = Dam

$\overline{AC}$ = Salmon migration

$\overline{AD}$ = Climate change

$\overline{AE}$ = Sediment

$\overline{BC}$ = Seaways

$\overline{BD}$ = In the mountains

$\overline{BE}$ = Turbidity

$\overline{CD}$ = Ice Age

$\overline{CE}$ = Acidity

$\overline{DE}$ = Eventually

Figure 79: In the city center

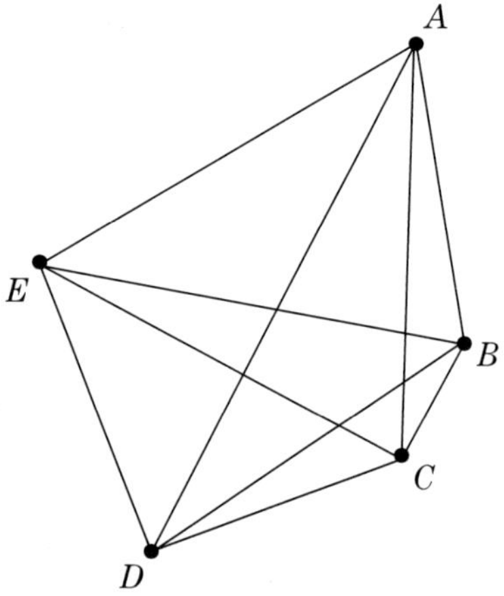

A = Time
B = Materials
C = Souls
D = Ideas
E = Places

$\overline{AB}$ = Estimated costs

$\overline{AC}$ = Immaterial

$\overline{AD}$ = Recess

$\overline{AE}$ = Settings

$\overline{BC}$ = Experiences

$\overline{BD}$ = Observations

$\overline{BE}$ = Natural resources

$\overline{CD}$ = Religions

$\overline{CE}$ = Hauntings

$\overline{DE}$ = Museums

***Figure 80*: Courses 101**

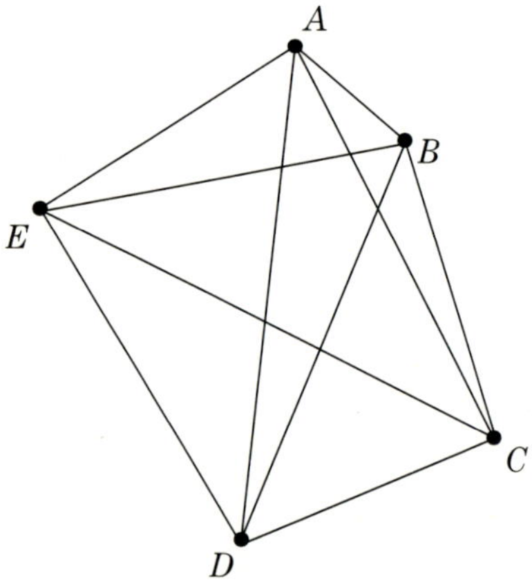

A = Appetizer

B = Soup

C = Salad

D = Main

E = Dessert

$\overline{AB}$ = Cup

$\overline{AC}$ = Side

$\overline{AD}$ = Tapas

$\overline{AE}$ = Fun-sized

$\overline{BC}$ = Gazpacho

$\overline{BD}$ = Stew

$\overline{BE}$ = Melted ice cream

$\overline{CD}$ = Cobb

$\overline{CE}$ = Jello

$\overline{DE}$ = Dinner at Grandma's

***Figure 81*: Some dreams are shallower than others**

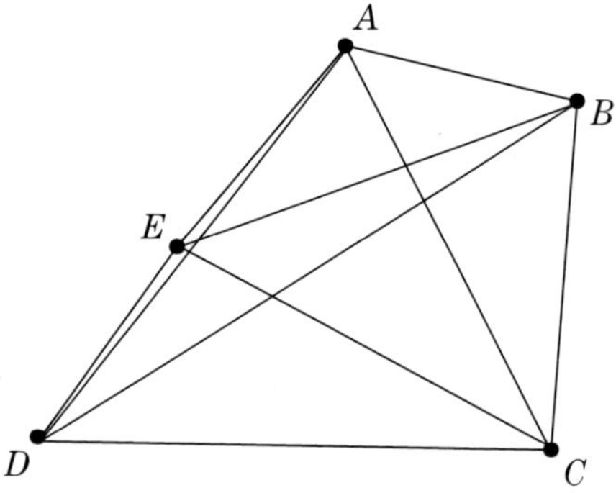

A = Vacation
B = Job
C = House
D = Boyfriend
E = Body

$\overline{AB}$ = Retirement

$\overline{AC}$ = Beach

$\overline{AD}$ = Solitude

$\overline{AE}$ = Free from petty oversights

$\overline{BC}$ = Mortgage payments

$\overline{BD}$ = Emotional chores

$\overline{BE}$ = Plastic surgery

$\overline{CD}$ = Slacker

$\overline{CE}$ = Real estate of the soul

$\overline{DE}$ = Underwhelming

Figure 82: How do you two know each other?

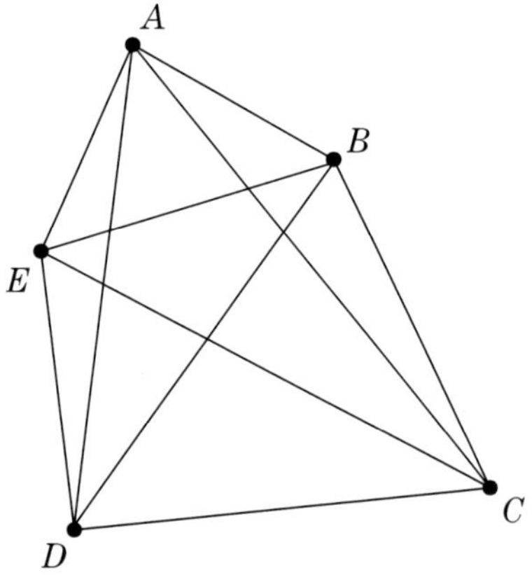

A = Love

B = Lust

C = Like

D = Know

E = Lose

$\overline{AB}$ = Hormones

$\overline{AC}$ = Good children

$\overline{AD}$ = Learning

$\overline{AE}$ = Widowhood

$\overline{BC}$ = Crushes

$\overline{BD}$ = Sex education

$\overline{BE}$ = Menopause

$\overline{CD}$ = Curiosity

$\overline{CE}$ = Friends who move away

$\overline{DE}$ = Fading memories

Figure 83: Contained populations

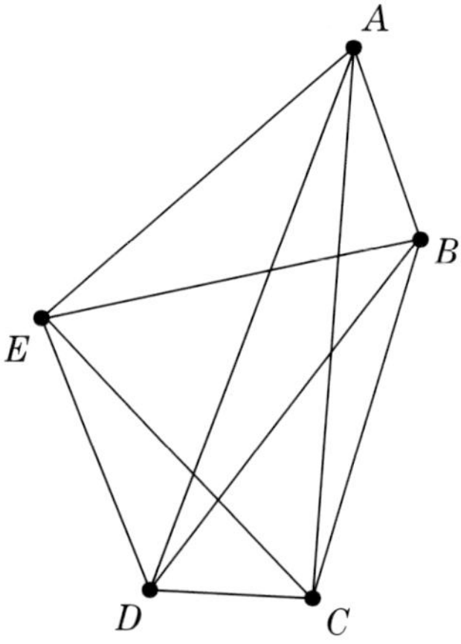

A = Jail
B = Airport
C = School
D = Office
E = Zoo

$\overline{AB}$ = Delayed flight

$\overline{AC}$ = Sixth grade

$\overline{AD}$ = Time cards

$\overline{AE}$ = Pet quarantine

$\overline{BC}$ = Flight simulator

$\overline{BD}$ = Customs

$\overline{BE}$ = Day before Thanksgiving

$\overline{CD}$ = Where bad kids go

$\overline{CE}$ = Veterinary internship

$\overline{DE}$ = Boss is away

Figure 84: Flavors of anxiety

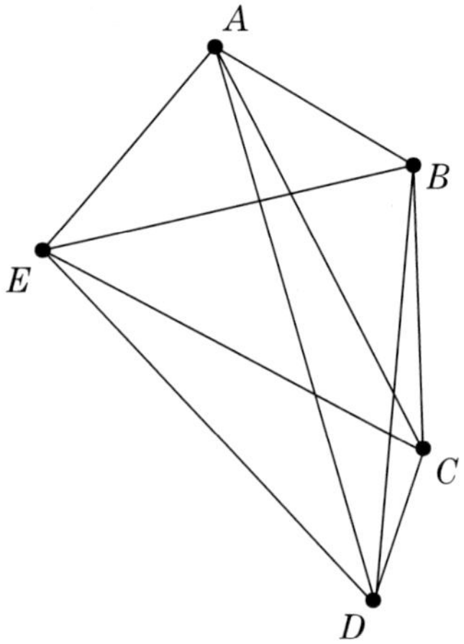

A = Performance
B = Obsessive
C = Panic
D = Post-traumatic
E = Social

$\overline{AB}$ = Actors

$\overline{AC}$ = Stage fright

$\overline{AD}$ = Laughing women

$\overline{AE}$ = False smiles

$\overline{BC}$ = Petrification

$\overline{BD}$ = It never stopped happening

$\overline{BE}$ = Paranoia

$\overline{CD}$ = Nights terrors

$\overline{CE}$ = Not attending the party

$\overline{DE}$ = Bad break-up

***Figure** 85*: **What we're reaching for today**

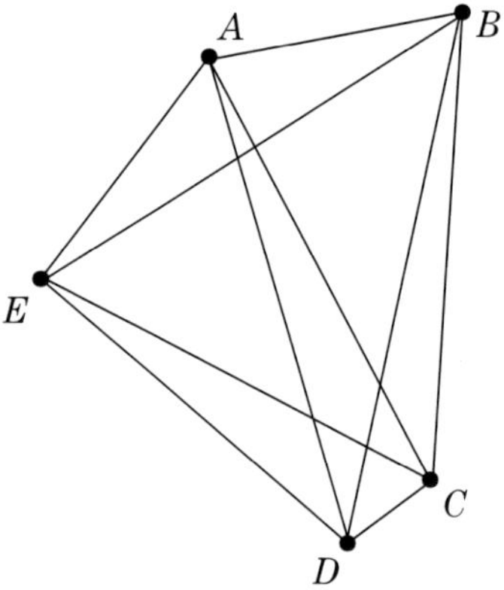

A = The moon
B = The finish line
C = His heart
D = The top shelf
E = My potential

$\overline{AB}$ = The marathon

$\overline{AC}$ = The one who got away

$\overline{AD}$ = The biggest awards

$\overline{AE}$ = To be someone else

$\overline{BC}$ = "I love you, too."

$\overline{BD}$ = Celebratory champagne

$\overline{BE}$ = Beyond the horizon

$\overline{CD}$ = Out of my league

$\overline{CE}$ = Really mattering to someone

$\overline{DE}$ = What could have been

***Figure 86*: We live in a melting greenhouse**

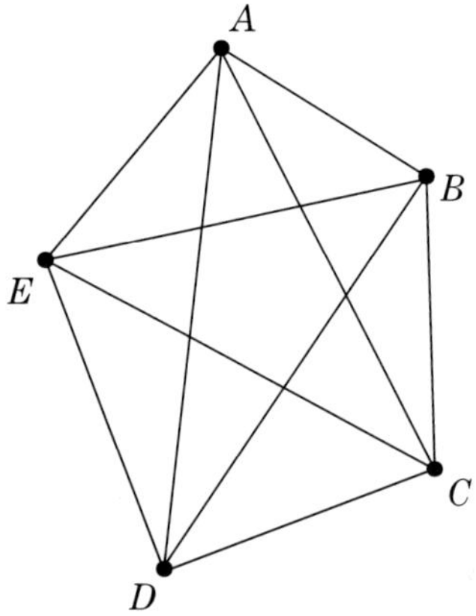

A = Extinction
B = Migration
C = Acidification
D = Liquefaction
E = Suffocation

$\overline{AB}$ = Nowhere to go

$\overline{AC}$ = Coral reef bleaching

$\overline{AD}$ = Melting glaciers

$\overline{AE}$ = Deforestation

$\overline{BC}$ = Oceanic dead zones

$\overline{BD}$ = Thawing permafrost

$\overline{BE}$ = Smog refugees

$\overline{CD}$ = Acid rain

$\overline{CE}$ = Chemical weapons

$\overline{DE}$ = Carbon monoxide poisoning

Figure 87: Fields of study II

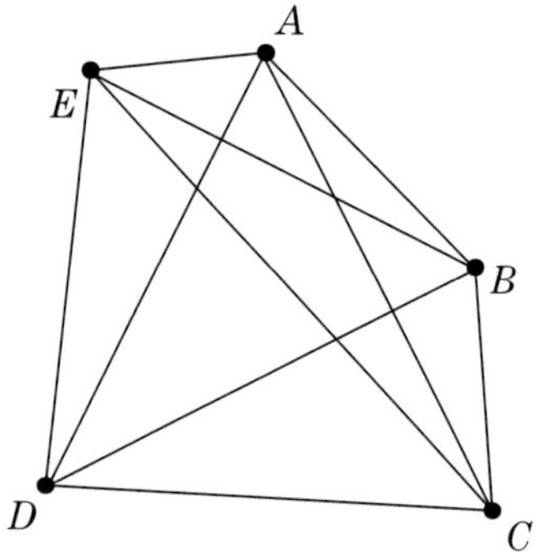

A = Technology
B = Mythology
C = Anthropology
D = Archeology
E = Psychology

$\overline{AB}$ = Prometheus

$\overline{AC}$ = Social media

$\overline{AD}$ = Carbon dating

$\overline{AE}$ = Skinner box

$\overline{BC}$ = Ruined temples

$\overline{BD}$ = Dragon or dinosaur?

$\overline{BE}$ = Jung

$\overline{CD}$ = Cave paintings

$\overline{CE}$ = Cultural mindsets

$\overline{DE}$ = Dredging up memories

***Figure** 88*: **Moving people**

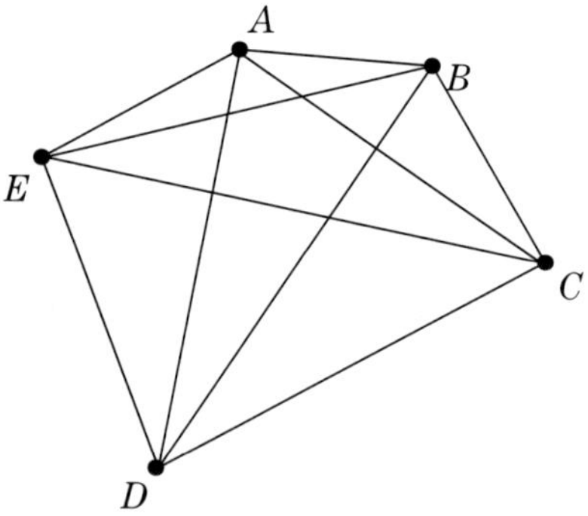

A = Roller coast
B = Bounce house
C = Ski lift
D = Movie theater
E = Cathedral

$\overline{AB}$ = Happy kids

$\overline{AC}$ = High winds

$\overline{AD}$ = Action movie

$\overline{AE}$ = Unfolding clergy scandal

$\overline{BC}$ = Jump to safety

$\overline{BD}$ = Rich kid's mansion wing

$\overline{BE}$ = Churchyard picnic

$\overline{CD}$ = Park City

$\overline{CE}$ = Rise up, don't fall

$\overline{DE}$ = IMAX

Figure 89: How relationships end

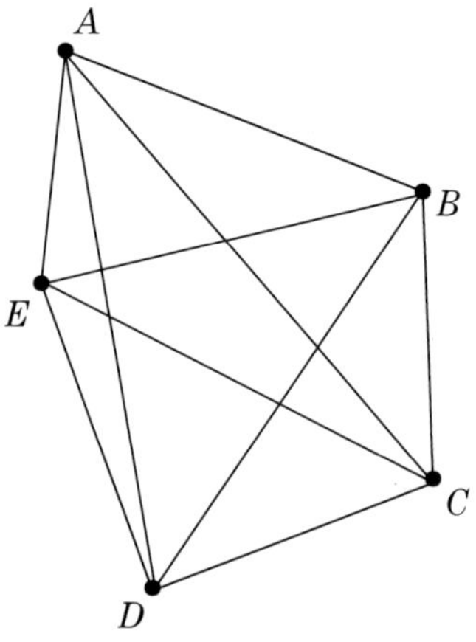

A = Kissing
B = Telling
C = Going
D = Staying
E = Silence

$\overline{AB}$ = Revenge porn

$\overline{AC}$ = One-night stands

$\overline{AD}$ = Marriage

$\overline{AE}$ = Abuse

$\overline{BC}$ = Dear John,

$\overline{BD}$ = Advocates

$\overline{BE}$ = Expired statute of limitations

$\overline{CD}$ = Hotels

$\overline{CE}$ = Ghosting

$\overline{DE}$ = Futile martyrdom

***Figure 90*: Powerful women**

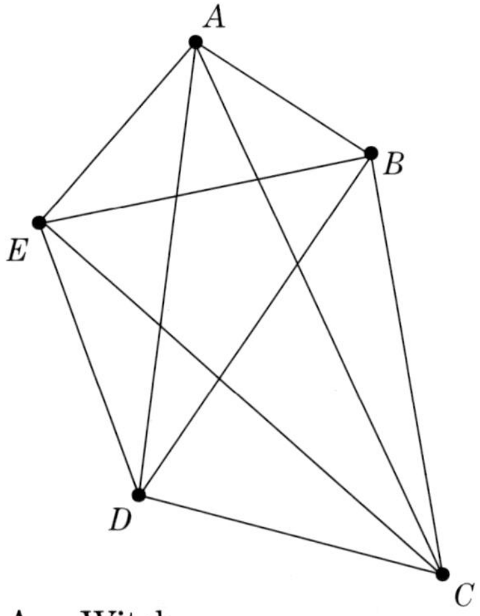

A = Witch

B = Gorgon

C = Succubus

D = Harpy

E = Siren

$\overline{AB}$ = Chief Executive Officer

$\overline{AC}$ = Chief Financial Officer

$\overline{AD}$ = Chief Compliance Officer

$\overline{AE}$ = Chief Technology Officer

$\overline{BC}$ = Chief Information Officer

$\overline{BD}$ = Chief Innovation Officer

$\overline{BE}$ = Chief Creative Officer

$\overline{CD}$ = Chief Efficiency Officer

$\overline{CE}$ = Chief Security Officer

$\overline{DE}$ = Chief Knowledge Officer

Figure 91: One one hand

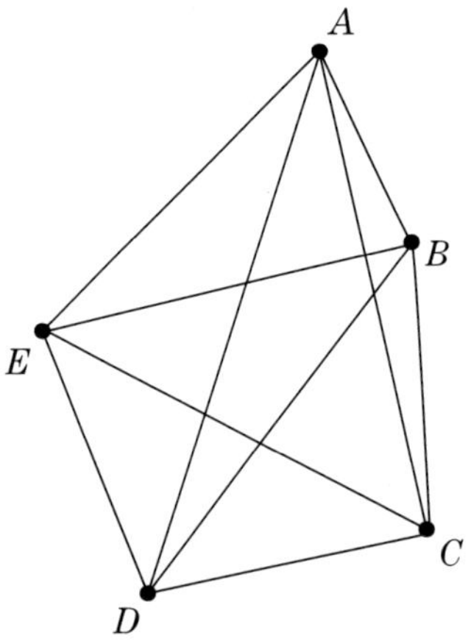

A = Thumb
B = Pointer
C = Middle
D = Ring
E = Pinky

$\overline{AB}$ = Okay

$\overline{AC}$ = Ready to flick

$\overline{AD}$ = Shocker

$\overline{AE}$ = Girl Scout promise

$\overline{BC}$ = Hopefully

$\overline{BD}$ = The bird

$\overline{BE}$ = In mittens

$\overline{CD}$ = West Side

$\overline{CE}$ = Fidgeting

$\overline{DE}$ = Ulnar claw

Figure 92: Ways to be good

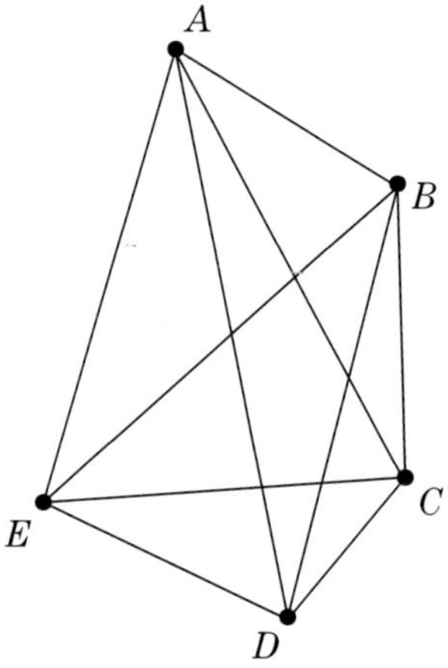

A = Damned
B = Pretty
C = Greater
D = Durable
E = Debatably

$\overline{AB}$ = Girls

$\overline{AC}$ = Faust

$\overline{AD}$ = Plastic

$\overline{AE}$ = Agnostics

$\overline{BC}$ = Jealousy

$\overline{BD}$ = Denim

$\overline{BE}$ = Striking

$\overline{CD}$ = Persistence

$\overline{CE}$ = Classic albums

$\overline{DE}$ = Condoms

Figure 93: Parasitic relationships

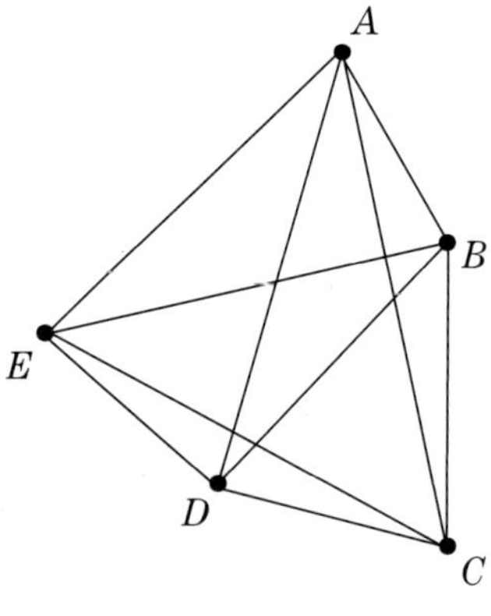

A = Intestinal
B = Corporate
C = Economic
D = Uterine
E = Political

$\overline{AB}$ = IBS drug patents

$\overline{AC}$ = Cholera

$\overline{AD}$ = Cramps

$\overline{AE}$ = Attack ads

$\overline{BC}$ = Full-time poverty wages

$\overline{BD}$ = Maternity leave policy

$\overline{BE}$ = Offshore tax havens

$\overline{CD}$ = Wage gap

$\overline{CE}$ = Pork barrel

$\overline{DE}$ = Handmaid's Tale

***Figure 94*: Identifying information**

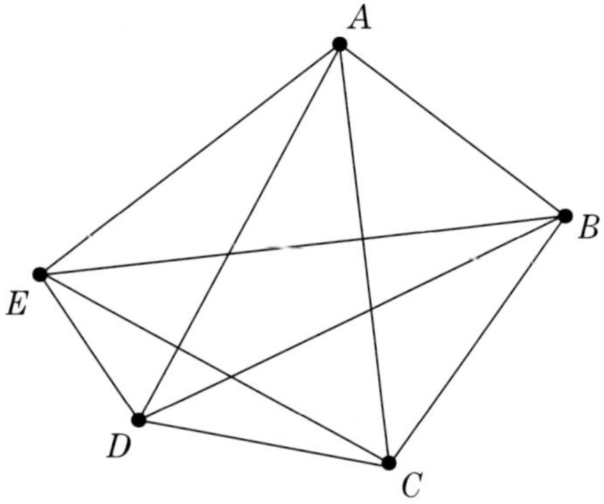

A = Name
B = Address
C = Birthday
D = Occupation
E = Citizenship

$\overline{AB}$ = Socioeconomic status

$\overline{AC}$ = Icing on the cake

$\overline{AD}$ = Business cards

$\overline{AE}$ = Can trigger xenophobia

$\overline{BC}$ = Security questions

$\overline{BD}$ = Quality of school districts

$\overline{BE}$ = For the lucky, a match

$\overline{CD}$ = Midwives

$\overline{CE}$ = America

$\overline{DE}$ = Skilled workers

***Figure 95*: How to belittle others**

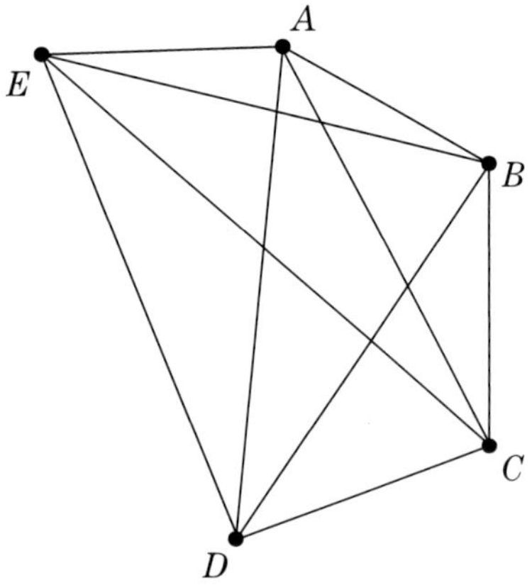

A = Ugly
B = Stupid
C = Dirty
D = Criminal
E = Promiscuous

$\overline{AB}$ = Assumptions

$\overline{AC}$ = Slurs

$\overline{AD}$ = Discrimination

$\overline{AE}$ = Nazi memes

$\overline{BC}$ = Bad apples

$\overline{BD}$ = Public violence

$\overline{BE}$ = Casual hate speech

$\overline{CD}$ = Administrative leave

$\overline{CE}$ = Planting evidence

$\overline{DE}$ = Jim Crow

Figure 96: Ghosts of time past

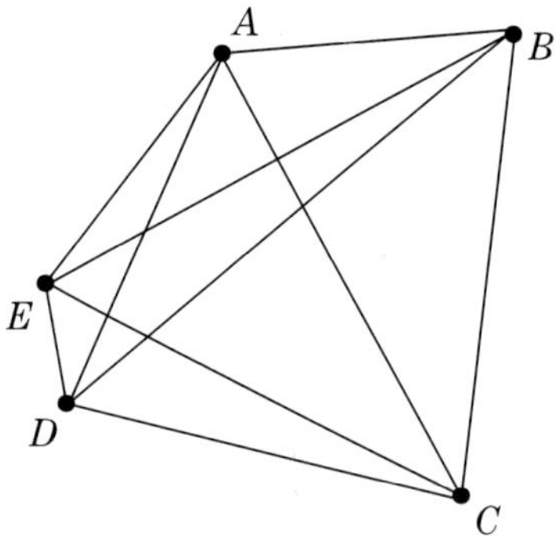

A = Mastodons
B = Gentlemen
C = Dodos
D = Rotary phones
E = Grandma

$\overline{AB}$ = Cavemen who draw on walls

$\overline{AC}$ = Not even in zoos

$\overline{AD}$ = Ivory handles

$\overline{AE}$ = She hunted him long ago

$\overline{BC}$ = So-called nice guys

$\overline{BD}$ = Grandpa asks for a date

$\overline{BE}$ = Chaperoned dates

$\overline{CD}$ = Stuffed in dust museums

$\overline{CE}$ = Not chosen to be Grandpa

$\overline{DE}$ = "Not with my arthritis."

***Figure 97*: How do you solve a problem like Maria?**

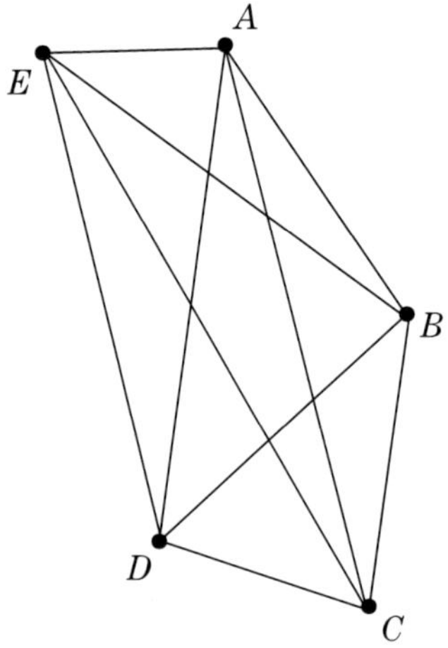

A = Exploit
B = Ignore
C = Enrich
D = Befriend
E = Erase

$\overline{AB}$ = Not tipping

$\overline{AC}$ = Employ

$\overline{AD}$ = Smalltalk with the help

$\overline{AE}$ = Downsizing

$\overline{BC}$ = Conservation

$\overline{BD}$ = Overlooking flaws

$\overline{BE}$ = Ghosting

$\overline{CD}$ = Commiseration

$\overline{CE}$ = Righting wrongs

$\overline{DE}$ = Forgiveness

***Figure 98*: What did you do at work today?**

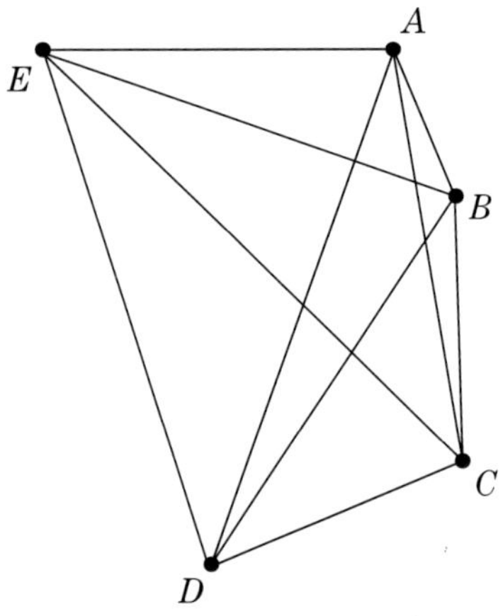

A = Organize

B = Strike

C = Toil

D = Submit

E = Complain

$\overline{AB}$ = Union reps

$\overline{AC}$ = Org charts

$\overline{AD}$ = Propaganda

$\overline{AE}$ = Letter-writing campaigns

$\overline{BC}$ = Picketing

$\overline{BD}$ = Reaching new deals

$\overline{BE}$ = "He hit me first."

$\overline{CD}$ = Subservience

$\overline{CE}$ = Venting to coworkers

$\overline{DE}$ = File suit

***Figure 99*: Retirement options**

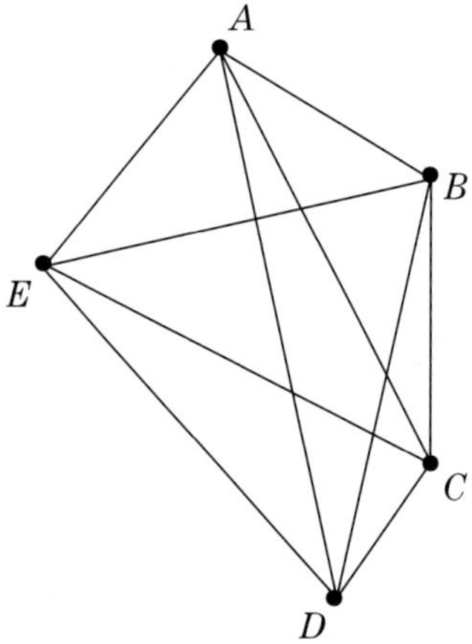

A = Quit
B = Travel
C = Charity
D = Putter
E = Ennui

$\overline{AB}$ = Lottery winners

$\overline{AC}$ = Scrooge before ghosts

$\overline{AD}$ = Career change

$\overline{AE}$ = Motivation

$\overline{BC}$ = Mission trips

$\overline{BD}$ = Golf trips

$\overline{BE}$ = Every place feels the same

$\overline{CD}$ = Easy volunteering

$\overline{CE}$ = Overwhelming problems

$\overline{DE}$ = Isolation

***Figure 100*: Personal assets**

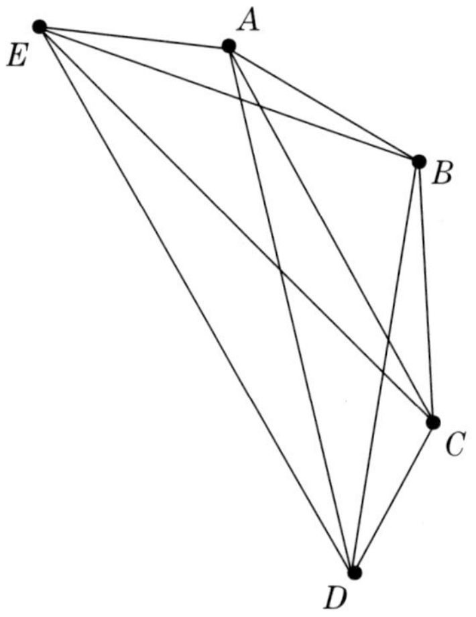

A = Charm
B = Wisdom
C = Wealth
D = Power
E = Health

$\overline{AB}$ = Parables

$\overline{AC}$ = Taste

$\overline{AD}$ = Charisma

$\overline{AE}$ = Moderation

$\overline{BC}$ = Knowing what enough is

$\overline{BD}$ = Restraint

$\overline{BE}$ = Daily apples

$\overline{CD}$ = Buying politicians

$\overline{CE}$ = Access to medicine

$\overline{DE}$ = Athletics

Figure 101: Discussable fears

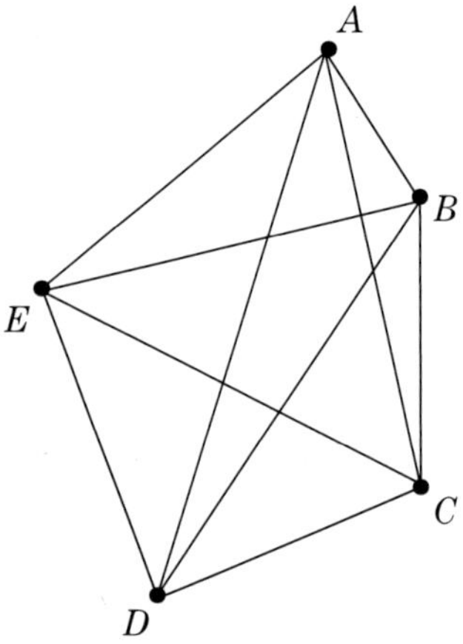

A = Clowns
B = Public speaking
C = Failure
D = Spiders
E = Death

$\overline{AB}$ = Time-share sales pitches

$\overline{AC}$ = Death at the rodeo

$\overline{AD}$ = Abandoned circus

$\overline{AE}$ = Single-car accident

$\overline{BC}$ = Um. Uh? Um.

$\overline{BD}$ = Arachnology conference

$\overline{BE}$ = Assassinated at the podium

$\overline{CD}$ = Webless

$\overline{CE}$ = Attempted murder

$\overline{DE}$ = Brown recluse

Figure 102: Aging out of here

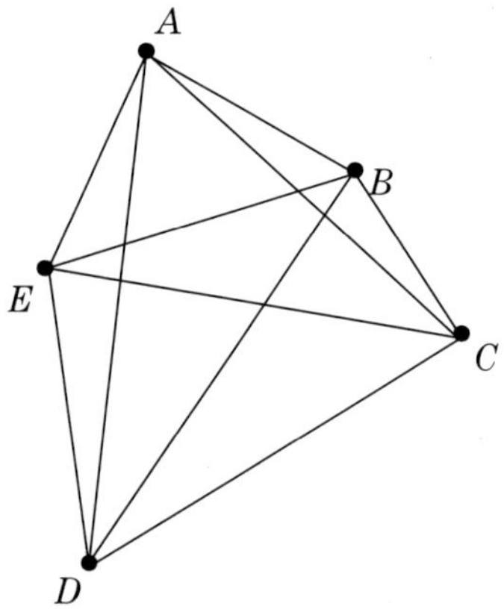

A = Infancy
B = Childhood
C = Adolescence
D = Adulthood
E = Decline

$\overline{AB}$ = Fragility

$\overline{AC}$ = Teen pregnancy

$\overline{AD}$ = First job

$\overline{AE}$ = Tragedy

$\overline{BC}$ = Puberty

$\overline{BD}$ = DINKs

$\overline{BE}$ = Vasectomy

$\overline{CD}$ = College

$\overline{CE}$ = Joining the rough crowd

$\overline{DE}$ = Hospice

Figure 103: What they call you behind your back

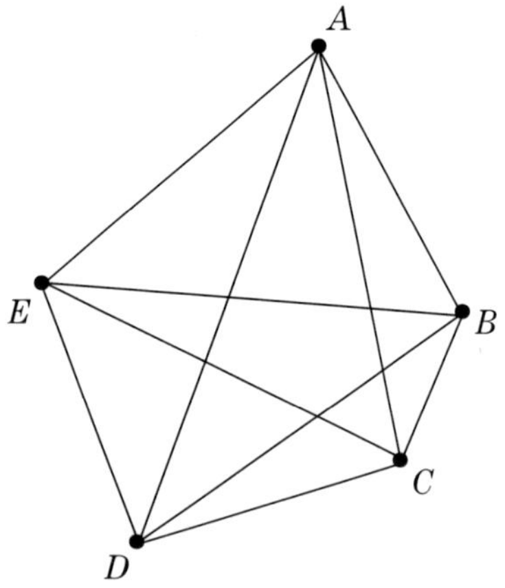

A = Difficult
B = Idiot
C = Genius
D = Sweetheart
E = Monster

$\overline{AB}$ = Baby

$\overline{AC}$ = Icon

$\overline{AD}$ = Tease

$\overline{AE}$ = Beowulf

$\overline{BC}$ = Savant

$\overline{BD}$ = Taken advantage of

$\overline{BE}$ = Sunbathing vampire

$\overline{CD}$ = Saint

$\overline{CE}$ = Master thief

$\overline{DE}$ = Abuser

Figure 104: Adventures in housewifery

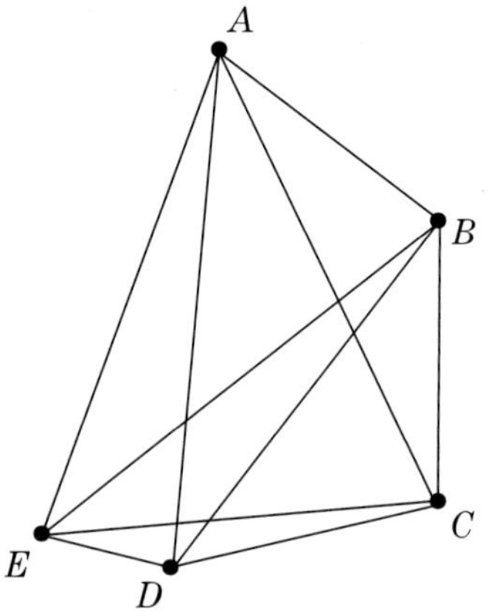

A = Cooking
B = Cleaning
C = Caring
D = Cursing
E = Curating

$\overline{AB}$ = Dishes

$\overline{AC}$ = Feeding

$\overline{AD}$ = Burned it again

$\overline{AE}$ = Meal planning

$\overline{BC}$ = Bathing

$\overline{BD}$ = Wiping

$\overline{BE}$ = De-cluttering

$\overline{CD}$ = Furious empathy

$\overline{CE}$ = Put on his favorite shows

$\overline{DE}$ = Shock art

***Figure 105*: Adjectives girls hide their rage underneath**

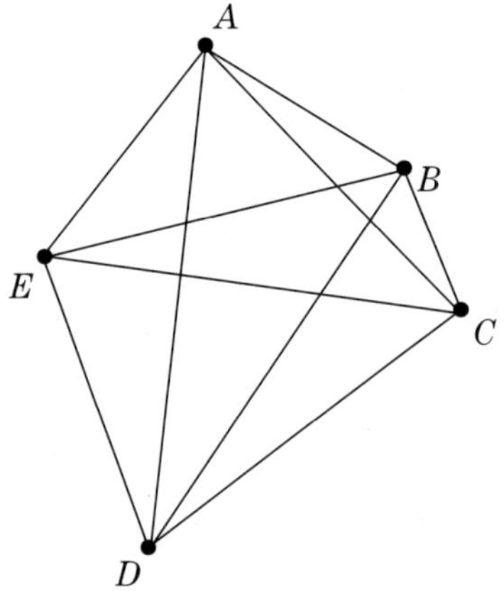

A = Chaste
B = Slender
C = Silent
D = Obedient
E = Diligent

$\overline{AB}$ = Hungry

$\overline{AC}$ = Repressed

$\overline{AD}$ = Resentful

$\overline{AE}$ = Studious

$\overline{BC}$ = Decorative

$\overline{BD}$ = Controlled

$\overline{BE}$ = Exhausted

$\overline{CD}$ = Seething

$\overline{CE}$ = Plotting

$\overline{DE}$ = Afraid

Figure 106: Pieces of you

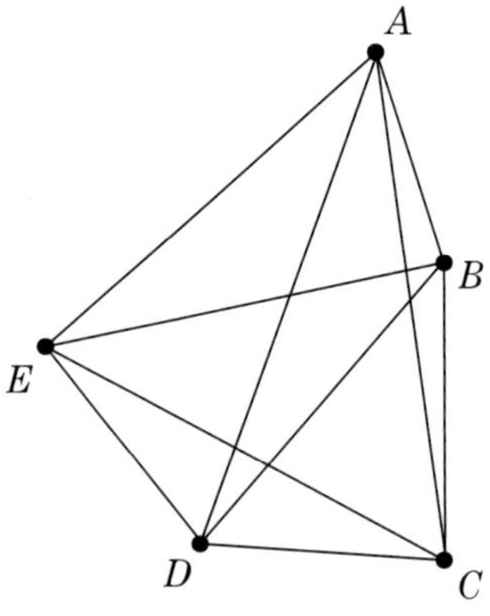

A = Liver
B = Eyes
C = Blood
D = Lung
E = Heart

$\overline{AB}$ = Jaundice

$\overline{AC}$ = Sausages

$\overline{AD}$ = Filters

$\overline{AE}$ = Atherosclerosis

$\overline{BC}$ = Broken capillaries

$\overline{BD}$ = Stung by smoke

$\overline{BE}$ = Love at first sight

$\overline{CD}$ = Clot

$\overline{CE}$ = Flow

$\overline{DE}$ = Pulmonary

***Figure 107*: Characters you've met and been**

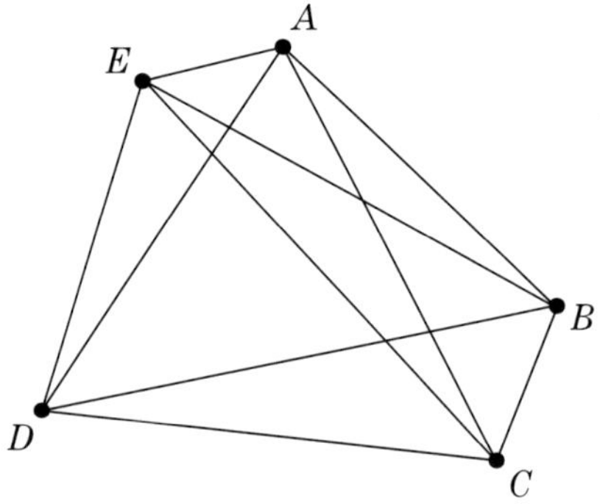

A = Hero
B = Villain
C = Sidekick
D = Narrator
E = Love interest

$\overline{AB}$ = Autobiographer

$\overline{AC}$ = Secretary

$\overline{AD}$ = Egotist

$\overline{AE}$ = White knight

$\overline{BC}$ = Backstabber

$\overline{BD}$ = Tale of revenge

$\overline{BE}$ = Vampire story

$\overline{CD}$ = Ishmael

$\overline{CE}$ = Overlooked

$\overline{DE}$ = Romance

***Figure 108*: Role call**

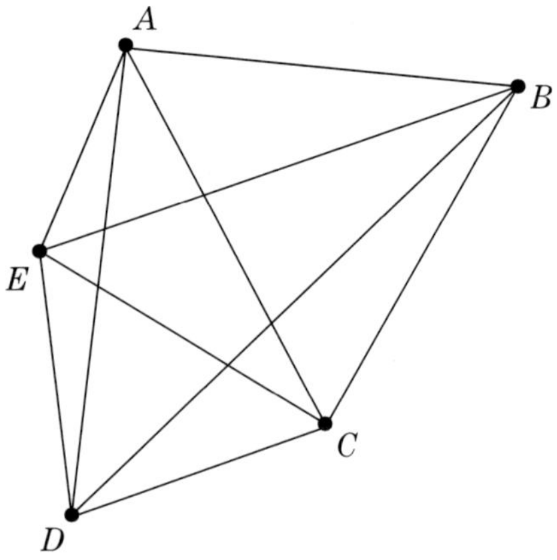

A = Nerds
B = Jocks
C = Preps
D = Goths
E = Jesus freaks

$\overline{AB}$ = Baseball statistics

$\overline{AC}$ = Debate

$\overline{AD}$ = Term papers on catacombs

$\overline{AE}$ = Bible study

$\overline{BC}$ = Golf

$\overline{BD}$ = Blood doping

$\overline{BE}$ = Thank you, God, for the win

$\overline{CD}$ = Updike stories

$\overline{CE}$ = Modest khakis

$\overline{DE}$ = Crucifixes

Figure 109: Thanks for that, Grandma

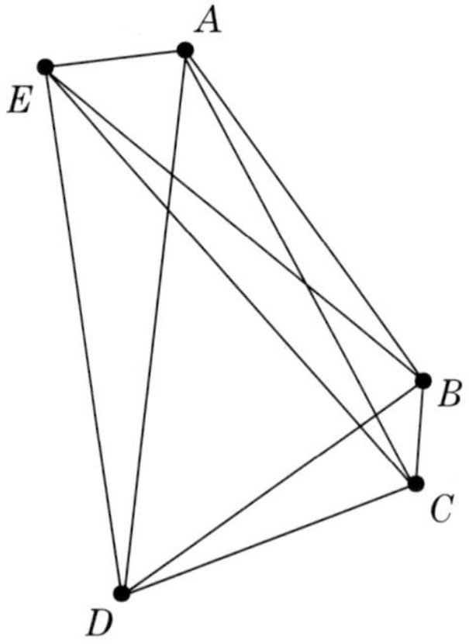

A = Dusty candy dishes
B = Yellowed doilies
C = House slippers
D = Prescription bottles
E = Stray cats

$\overline{AB}$ = Few visitors

$\overline{AC}$ = Ghostly quiet

$\overline{AD}$ = Placebos

$\overline{AE}$ = Treats left out

$\overline{BC}$ = Soft relics

$\overline{BD}$ = Melancholy centerpieces

$\overline{BE}$ = Ammonia aroma

$\overline{CD}$ = Arthritis

$\overline{CE}$ = Scratched silk

$\overline{DE}$ = Take with food

***Figure 110*: Not that kind of girl?**

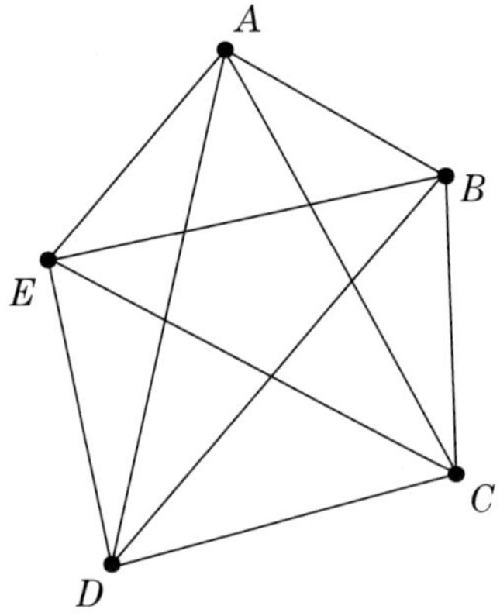

A = Virgin
B = Whore
C = Mother
D = Crone
E = Wife

$\overline{AB}$ = Can't win

$\overline{AC}$ = Mary

$\overline{AD}$ = Spinster

$\overline{AE}$ = Beard

$\overline{BC}$ = Insult

$\overline{BD}$ = Survivor

$\overline{BE}$ = Cuckolding

$\overline{CD}$ = Grandma

$\overline{CE}$ = Epitaph

$\overline{DE}$ = He still loves her

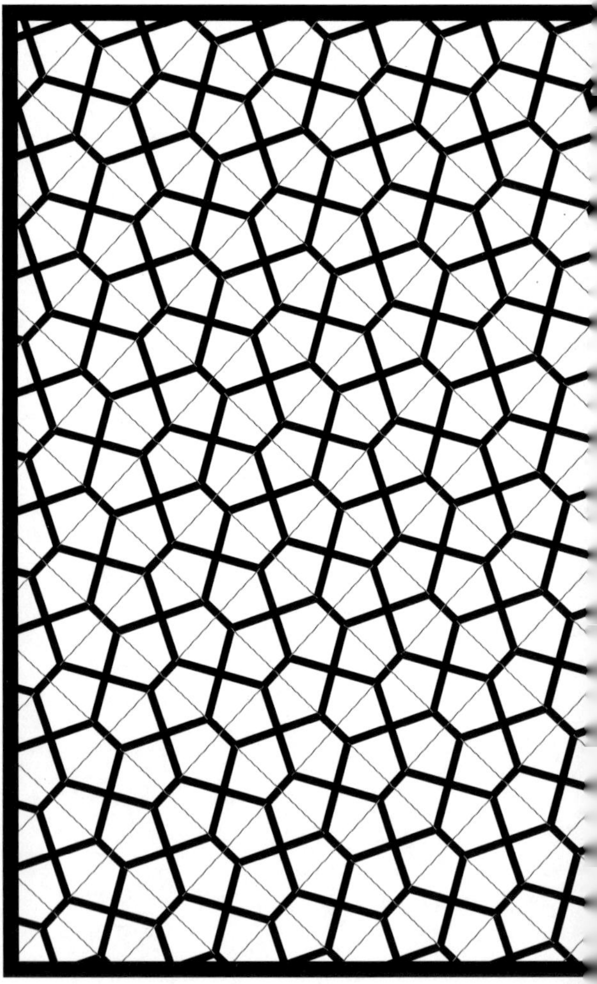

AETUI is the 36th book from Inside the Castle or ITC-036.

ISBN 978-1-7352901-7-1

21 22 23 24 25 26 27 28 29 30 SH SH 8 7 6 5 4 3

DESIGNED AND EDITED BY JOHN TREFRY:

Significant design inspiration is derived from the 1963 book, *Schaum's Outline of Plane Geometry With Coordinate Geometry*, by Dr. Barnett Rich, former chair of the mathematics department at the Brooklyn Technical High School in New York City. Dr. Rich's book appears in Paul Verhoeven's 1987 tragic masterpiece *Robocop* being read by a nervous gas station attendant who is most likely a surrogate for Verhoeven himself.

THE TYPEFACE:

The typeface whose many forms are liberally employed in *Plane Geometry* and *AETUI* is Century Expanded — a variation on the original Century Expanded typeface of the early 1900s — that was specifically designed for the advent of cold type processes, or phototypesetting, precipitating the obsolescence of metal type in the early 1960s. Additionally, Bodoni is used in certain areas of the book to bring it into conformance with the aesthetic flourishes of *Plane Geometry*.

Catalog

If you are interested in a list of INSIDE THE CASTLE titles in print you may visit www.insidethecastle.org.

Inside the Castle is located in Lawrence, Kansas

Made in United States
North Haven, CT
10 November 2026

22088879R00136